SpringerBriefs in Computer Science

SpringerBriefs present concise summaries of cutting-edge research and practical applications across a wide spectrum of fields. Featuring compact volumes of 50 to 125 pages, the series covers a range of content from professional to academic

Typical topics might include:

- A timely report of state-of-the art analytical techniques
- A bridge between new research results, as published in journal articles, and a contextual literature review
- A snapshot of a hot or emerging topic
- An in-depth case study or clinical example
- A presentation of core concepts that students must understand in order to make independent contributions

Briefs allow authors to present their ideas and readers to absorb them with minimal time investment. Briefs will be published as part of Springer's eBook collection, with millions of users worldwide. In addition, Briefs will be available for individual print and electronic purchase. Briefs are characterized by fast, global electronic dissemination, standard publishing contracts, easy-to-use manuscript preparation and formatting guidelines, and expedited production schedules. We aim for publication 8-12 weeks after acceptance. Both solicited and unsolicited manuscripts are considered for publication in this series.

More information about this series at http://www.springer.com/series/10028

Philippe De Ryck • Lieven Desmet
Frank Piessens • Martin Johns

Primer on Client-Side Web Security

 Springer

Philippe De Ryck
iMinds-DistriNet
KU Leuven
Heverlee
Belgium

Lieven Desmet
iMinds-DistriNet
KU Leuven
Heverlee
Belgium

Frank Piessens
iMinds-DistriNet
KU Leuven
Heverlee
Belgium

Martin Johns
SAP Research
Karlsruhe
Germany

ISSN 2191-5768   ISSN 2191-5776 (electronic)
ISBN 978-3-319-12225-0   ISBN 978-3-319-12226-7 (eBook)
DOI 10.1007/978-3-319-12226-7
Springer Cham Heidelberg New York Dordrecht London

Library of Congress Control Number: 2014953777

Printed on acid-free paper

Springer is part of Springer Science+Business Media (www.springer.com)

Preface

Have you ever wondered why all of a sudden, normal users start posting strange messages on social networks? How wireless routers can be controlled remotely? Why eBay accounts could be hijacked with a single HTTP request? Or why a news Web site suddenly shows a page from the Syrian Electronic Army? All of these incidents were possible due to attackers controlling some code within the victim's browser, a result of the current state of practice in Web security, which is less than stellar. As security researchers, we are concerned by the large gap between the state of practice and the currently available security technologies, which are often inspired by security research. In an effort to improve this situation, we have written this book, which gives a detailed view on the client-side Web security landscape. We explicitly focus on client-side security vulnerabilities, which are exploited from within a browser or explicitly target the browser, because they generally receive less attention compared to their server-side counterparts. In total, we cover 13 attacks, for which we give a detailed description, an overview of traditional mitigation techniques, and current state-of-the-art research. For each attack, we also describe the current state of practice in Web applications, and define the best practices to defend against these attacks in the modern age.

We have written this book with several target audiences in mind. It offers *students, teachers, and trainers* an introduction into the field of client-side Web security, with an extensive reference list for learning more about each topic. The best practices can be translated into teaching material for secure software development courses. The book helps *junior researchers* to quickly get up to speed in the field, and offers an overview of the current state-of-the-art for *experienced researchers*, who are looking for new opportunities to explore. Finally, *developers and security practitioners* get an overview of the current state of practice, and the upcoming state-of-the-art technologies. They should use the best practices in the book to improve the state of practice, which is beneficial for all users on the Web.

This book grew from our experience as security researchers[1] working on Web security, with a strong focus on client-side Web security topics such as cross-site request forgery, cross-site scripting, session management problems, and click-jacking. We also actively participate in European Web security projects, such as STREWS[2], WebSand[3], and NESSoS[4], and collaborate with the W3C and IETF standardization committees, further expanding our view on the current state of practice, state-of-the-art, and best practices.

We would like to explicitly acknowledge the support of the Agency for Innovation by Science and Technology (IWT), the STREWS project, where a preliminary version of this book was written as a first deliverable, and the IWT-SBO project SPION[5], which provided valuable insights in the privacy and security concerns of contemporary Web applications.

---

[1] Philippe De Ryck, Lieven Desmet, and Frank Piessens are affiliated with the *iMinds-DistriNet* research group at *KU Leuven University (Belgium)*, and Martin Johns is affiliated with *SAP Research (Germany)*.

[2] https://www.strews.eu/.

[3] https://www.websand.eu/.

[4] http://www.nessos-project.eu/.

[5] http://www.spion.me/.

Contents

Chapter 1
The Relevance of Client-Side Web Security

Google [15], LinkedIn [16], Adobe [7], Yahoo [4], eBay [17], Nintendo [11], Last-Pass [10], Vodafone [12], Target [18], Reuters [9]—there may not seem to be an apparent commonality between these companies, but they have all been victims of Web-based attacks resulting in the compromise of customer accounts, large-scale theft of customer information, or embarrassing defacements of their Web sites. The list includes ten prominent companies which are well aware of the dangers of the Web, and they are only the tip of the iceberg. A report about Web security in 2013 lists 253 data breaches [22], good for exposing a total of 552 million identities, and reports an astonishing 568,700 Web attacks blocked *per day*. Statistics show that cybercrime affects 378 million victims per year or 12 victims per second. Financially, the direct global loss induced by cybercrime amounts to $ 113 billion in a single year, enough to host the London Olympics about 10 times over [21].

The averse effects of these Web attacks are often underestimated, both for companies and for individuals. Companies that have become victims of a data breach or defacement not only suffer from business disruptions but also face investigations and potential lawsuits. Additionally, the ensuing reputation damage can cause long-term harmful effects, with customers leaving and shareholders losing confidence. Even worse, a continuous stream of security breaches can cause a loss of confidence in online services among the general population, severely hurting the online retail economy, e-government, and e-health services.

A 2013 survey [23] reports that 70 % of surveyed Internet users are concerned that their personal information is not kept secure by Web sites, resulting in adapted behavior, as 34 % of the users are less likely to give personal information on Web sites. And indeed, security breaches often cause significant collateral damage to individual users. For example, a stolen database of personal information often contains users' email addresses and even recoverable passwords. If the same credentials are used for the email account, the user can lose control over this account, as well as over all accounts that are associated with that email address. Even worse, the stolen information can be used to commit identity theft, resulting in fraudulent costs being attributed to the victim, instead of the perpetrator.

In other cases, the Web attack is only used as a stepping stone towards the compromise of a larger target. For example, Belgacom, a Belgian telco also running

© Philippe De Ryck, Lieven Desmet, Frank Piessens, Martin Johns 2014
P. De Ryck et al., *Primer on Client-Side Web Security*,
SpringerBriefs in Computer Science, DOI 10.1007/978-3-319-12226-7_1

infrastructure in Africa, was targeted by the British intelligence service GCHQ [8] through a Web attack. The attackers faked a social network application to serve malware to a Belgacom engineer, allowing the attackers to further infiltrate the Belgacom infrastructure. Another example is the 2010 compromise of *apache.org*, where a number of Web vulnerabilities eventually led to the compromise of the machine holding the code repositories [5].

With cybercrime as a billion-dollar business, the Web is in a dire situation. Web security is more important than ever, today and in the future. Before we start discussing attackers, problems, and their countermeasures, we take a closer look at how the Web came to be the way it is today, and why client-side Web security, the topic of this book, has become so popular.

1.1 The Web at a Glance

The World Wide Web started out as a distributed hypertext system, where documents hosted on networked computers contain references to other documents, hosted on different networked computers. These documents can be retrieved using a browser, dedicated client software for viewing hypertext documents, and following hyperlinks embedded within the text of these documents.

In order to make such a distributed hypertext system work, three fundamental agreements (standards) are necessary:

1. **Resource Identifiers** URIs [2] (originally called URLs) provide universally dereferenceable identifiers for resources on the network.
2. **Transfer Protocol** HTTP [3] (Hypertext Transfer Protocol) is a universally supported transfer protocol that, in its bare essence, provides a simple mechanism to retrieve a resource across the network and to submit form data from a browser to a Web server. HTTP is a request/response-based client–server protocol: The browser will send a request to a server that (a) identifies the resource, (b) identifies the media types of representations that the client is willing to consume in response (e.g., plain text or HTML, GIF or PNG), and (c) potentially is characterized as a form submission. The server responds with a resource representation that fulfills these constraints.
3. **Content Format** HTML [1] is a broadly implemented format for content, and started as a simple and declarative markup language. The early versions already included an anchor element that permits embedding hyperlinks to resources identified by URIs within the text. Based on plain text with embedded "tags," this markup language can be written in a simple text editor and remarkably, it is often written in code editors to this day, approximately 20 years later.

Notably, these three basic agreements are loosely coupled: While the URIs we use, most frequently identify resources retrieved through HTTP, URIs can also be used to identify resources retrieved through other protocols (early on, FTP was frequently used to serve resources on the Web, and indeed, Web browsers included implementations of FTP, Gopher, and a number of other protocols). HTTP can be used to

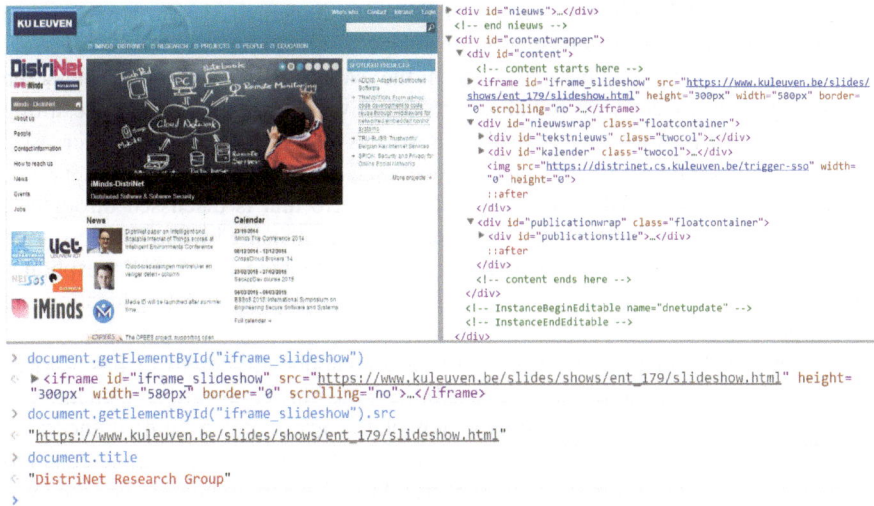

Fig. 1.1 An illustration of a rendered HTML page (*top left*) and its HTML code (*top right*) which is accessible from JavaScript (*bottom*) through the DOM interface

retrieve URI-identifiable resources in just about any format that is represented in bits and bytes, such as HTML documents, images, PDF documents, or scripts (i.e., executable program segments generally written in an interpretable scripting language such as JavaScript). Similarly, the concept of hyperlinks exists in formats beyond HTML: PDF, and even Word documents might permit these. Nevertheless, the Web's basic fabric is built on universal support for URIs, HTTP, and HTML.

The early Web was noninteractive, declarative, and stateless on the client side. While these properties, in all their simplicity, enabled the World Wide Web, they were insufficient to meet the demand for the rich application that the Web has become today. Without a doubt, JavaScript is the single most influential Web technology in the evolution of the Web. Initially intended to manipulate Web pages within the browser through the DOM or document object model (illustrated in Fig. 1.1), JavaScript quickly proved much more powerful, making it the de facto client-side programming language of the Web. One of the driving factors behind the success of JavaScript is AJAX, an asynchronous Web development technique based on JavaScript and XML, allowing Web pages to store and retrieve information in the background, updating the HTML of the Web page on the fly. Many modern Web applications still depend on this technique, albeit that XML has been replaced with the JavaScript Object Notation (JSON), a JavaScript-based format. A second technological upgrade consists of the rich content types available on the Web today. Modern Web content is no longer limited to HTML and images, as modern browsers also support several audio and video formats, XML-based languages for defining images (SVG) and scientific data (MathML), and advanced styling information for HTML documents (CSS). Finally, the third essential component for a rich application platform is client-side state within the browser, initially by means of cookies and later as full-fledged storage capabilities in databases or virtual file systems.

These three major changes on top of the basic hypertext system have sparked the shift from a one-way information exchange to a bidirectional read/write Web, also known as Web 2.0. This new stage in the evolution of the Web combines the technological advances with social aspects of actively participating users, resulting in dynamic applications, that actively improve as their number of users increases. Well-known examples are Wikipedia, Facebook, and the many Google services, which inspired the social network example scenario that is discussed throughout this book (introduced below). The social effect even intensified when people started carrying always-on, always-connected devices everywhere they go. Smartphones enable instant and continuous access to information, further stimulating location and context-aware social services.

Further development of the browser towards an application platform has resulted in a paradigm shift, where more responsibilities are pushed towards the client. The client component is no longer simply a view on the application running in the back-end but has become the application, which interacts with a light, storage-centered back-end application through rich, RESTful APIs. This "appification" of the Web is further stimulated by the rise of mobile devices, with their restricted operating systems, and vendor-controlled application stores. Not only is the majority of applications offered in today's application stores based on Web technology [13], but recent standardization efforts [6] provide the necessary APIs to build Web applications that can interact with the underlying device, making them indistinguishable from native applications.

The Web has known several evolutionary steps, which have transformed the static server-side content into dynamic server-side applications, and have transformed dull page-viewing browsers into execution platforms running highly dynamic and powerful Web applications. We observe a similar trend in the evolution of Web security, resulting in a shift from server-side to client-side Web security.

---

1 **Example Scenario: A Social Networking Application**
2 To better illustrate the security challenges in modern Web applications, we
3 introduce an example application that will be discussed throughout this book.
4 A social networking application serves as a perfect example, because it results
5 in an execution context with multiple stakeholders, with varying trust levels.
6 Additionally, the immense popularity of social networking sites makes the
7 example recognizable without the need for describing every feature in detail.
8 Our example application is aptly named *Our Social Network* and is virtually
9 hosted on *www.oursocialnetwork.com*. Figure 1.2 shows a conceptual overview
10 of the application, with four main components, each with a different trust level:
11 • The main context of the application responsible for embedding additional
12   components.
13 • The timeline, where users can post messages to their contacts and view
14   messages posted by their contacts. The timeline is shown in the middle.

Fig. 1.2 A conceptual depiction of our social networking example application. The application consists of several components, with varying trust levels, all assigned a different background color

- Commercial spaces, which users can add to their profile, to keep up-to-date with the latest business news. These commercial spaces can be purchased by business owners, and are hosted by *Our Social Network*. The commercial spaces are shown on the right.
- Third-party gadgets, which offer additional functionality, such as weather information or small games. These third-party gadgets are not provided nor hosted by *Our Social Network*, and are shown on the left.

Our example application, and every modern Web application, faces several security challenges. First, the social network needs to ensure that its traffic is protected against eavesdroppers and malicious intermediaries on the network. Second, the basic functionality offered by the social network depends on external JavaScript libraries that enable easy development of a responsive user interface, for example JQuery or AngularJS. Third, the application needs to integrate third-party gadgets into its main page but also likes to retain control over the behavior of the page, a challenge with currently available state of practice mechanisms. Finally, the social network offers commercial spaces to businesses as part of the *oursocialnetwork.com* domain. Therefore, the social network is not only responsible for the content but also needs to ensure that competitors do not influence each other, thereby causing reputation damage to the social network.

Table 1.1 The OWASP Top Ten Project [24] lists the most critical Web application security flaws. The *gray*-colored rows are relevant for client-side Web security and will be covered in this book

1	Injection
2	Broken Authentication and Session Management
3	Cross-Site Scripting (XSS)
4	Insecure Direct Object References
5	Security Misconfiguration
6	Sensitive Data Exposure
7	Missing Function Level Access Control
8	Cross-Site Request Forgery
9	Using Components with Known Vulnerabilities
10	Unvalidated Redirects and Forwards

1.2   Client-Side Web Security

The security landscape in the early Web was vastly different from what we see today. Attackers focused on server-side services, attempting to exploit the services or gain control over the server machine. Well-known examples of such attacks are SQL injection, command injection, or the exploitation of buffer overflow vulnerabilities within server software. As the functionalities of Web services grew, attackers started targeting client machines, aiming at exploiting client-side vulnerabilities to install malware, for example, to gain unauthorized access to the victim's bank accounts and other personal information. With the increasing security of browsers, the focus has shifted more towards the "weaker" Web vulnerabilities. Attacks such as cross-site scripting and cross-site request forgery use the browser as a means to carry out actions on the server, in the name of the victim.

A perfect illustration of the attacker's shift from server-side services towards the client side are two industry-driven surveys of the most important security vulnerabilities. Both the OWASP Top Ten Project [24] and the CWE/SANS Top 25 Most Dangerous Software Errors [14] include the typical server-side vulnerabilities, such as SQL injection and command injection, but also have allocated approximately one third of the slots to client-side security problems (shown in Tables 1.1 and 1.2).

Naturally, when the attackers' focus shifts towards the client, the countermeasures and security policies evolve as well. This evolution closely aligns with the evolutionary steps of the Web. The first security policies were static, encoded as default behavior in the browser, with the same rules for every Web application. Two examples of such static policies are the Same-Origin policy, and the same-origin behavior of the XMLHttpRequest object, two policies, which will be explained in the coming chapters. Next come the dynamic security policies, mainly enforced at the server side. Typical examples are token-based or request header-based protections against cross-site request forgery, or validation-based protections against cross-site scripting. These security policies line up with the rise of dynamic Web applications with a

Table 1.2 The CWE/SANS Top 25 Most Dangerous Software Errors [14] lists the most widespread and critical errors that lead to serious vulnerabilities. The *gray*-colored rows are relevant for client-side Web security and will be covered in this book

1	Improper Neutralization of Special Elements used in an SQL Command
2	Improper Neutralization of Special Elements used in an OS Command
3	Buffer Copy without Checking Size of Input ('Classic Buffer Overflow')
4	Improper Neutralization of Input During Web Page Generation ('Cross-site Scripting')
5	Missing Authentication for Critical Function
6	Missing Authorization
7	Use of Hard-coded Credentials
8	Missing Encryption of Sensitive Data
9	Unrestricted Upload of File with Dangerous Type
10	Reliance on Untrusted Inputs in a Security Decision
11	Execution with Unnecessary Privileges
12	Cross-Site Request Forgery (CSRF)
13	Improper Limitation of a Pathname to a Restricted Directory ('Path Traversal')
14	Download of Code Without Integrity Check
15	Incorrect Authorization
16	Inclusion of Functionality from Untrusted Control Sphere
17	Incorrect Permission Assignment for Critical Resource
18	Use of Potentially Dangerous Function
19	Use of a Broken or Risky Cryptographic Algorithm
20	Incorrect Calculation of Buffer Size
21	Improper Restriction of Excessive Authentication Attempts
22	URL Redirection to Untrusted Site ('Open Redirect')
23	Uncontrolled Format String
24	Integer Overflow or Wraparound
25	Use of a One-Way Hash without a Salt

server-side processing component. The next wave of security mechanisms takes advantage of the browser transformation into an application platform. Modern security policies are enforced at the client side but driven by the Web application at the server side. Prime examples are Content Security Policy [20], which is an application-specific policy enforced by the browser. Similarly, the recent work on Entry Points [19] proposes client-enforced protection against cross-site request forgery attacks. These server-driven, client-enforced policies are often used in a layered defense strategy, where both the client and the server enforce a security policy, hoping to stop an attacker that manages to circumvent one of the security measures.

This evolution where the client becomes the center of gravity for enforcing advanced security policies is exactly why this book focuses on client-side Web security. Client-side security policies are crucial for securing the Web in the future, but, as you will discover in this book, several challenges lie ahead of us.

1.3 Purpose of this Book

This book extensively covers the broad field of client-side Web security, in all its aspects, in order to help understand and position client-side Web security in the story of the Web. To this purpose, the book briefly covers the history of the Web, along with its fundamental building blocks and most recent evolutions. Based on a commonly used set of threat models, we investigate 13 different attacks, grouped into five chapters, based on their methods and impact. This book does not only cover the state-of-the-art technologies emerging from research and standardization activities, but also provides valuable insights into the current state of practice.

To be able to offer relevant information on the current state of practice, we have performed a large-scale study of the Alexa top 10,000 sites. We have trawled these Web sites, in total good for 4,185,227 requests, looking for deployments of well-known and recently introduced mitigation techniques. Based on these results, we can give an up-to-date view on the adoption rate of certain mitigation techniques, and show how even the most recent security technologies are already being adopted across the Web.

While the book is relevant for anyone aiming to learn about Web security and client-side countermeasures, the content is specifically tailored towards the following target audiences:

- **Students, Teachers, and Trainers**: Web development and Web security have become an indispensable part of academic computer science curricula and professional training programs. This book is ideally suited for Web security courses, as it provides the necessary background information, covers the different capabilities of attackers on the Web, and continues with a broad coverage of the Web's security problems and their countermeasures. The grouping of the attacks into chapters allows teachers and trainers to focus on the desired topics.
- **Researchers**: There is no lack of high-quality research on a wide variety of Web security topics, but being researchers ourselves, we noticed that it is hard to see the big picture. Therefore, we wrote this book to provide the big picture of the field of client-side Web security, covering both the attacks and the mitigation techniques. For every security problem, we describe the current state of practice as well as the latest research. The numerous citations make this book a timely reference work for both starting and experienced researchers, interested in discovering the current state-of-the-art research and the challenges that lie ahead.
- **Developers and Security Practitioners**: As you will learn from this book, many countermeasures depend on explicit developer action to ensure that Web applications are secured appropriately. Keeping up-to-date with all latest developments in the field of Web security is a daunting task. This book targets Web developers and security practitioners not only by offering an overview of the current Web security problems and their countermeasures but also by discussing the current state of practice in securing Web applications, as well as a set of best practices to secure a Web application.

References

1. Berjon, R., Faulkner, S., Leithead, T., Navara, E.D., O'Connor, E., Pfeiffer, S., Hickson, I.: HTML 5.1 specification. W3C Working Draft (2014)
2. Berners-Lee, T., Fielding, R.T., Masinter, L.: Uniform Resource Identifier (URI): generic syntax. RFC Internet Standard (RFC 3986) (2005)
3. Fielding, R., Gettys, J., Mogul, J., Frystyk, H., Masinter, L., Leach, P., Berners-Lee, T.: Hypertext Transfer Protocol – HTTP/1.1. RFC 2616 (1999)
4. Fitzgerald, D.: Yahoo passwords stolen in latest data breach. http://online.wsj.com/news /articles/SB10001424052702304373804577522613740363638 (2012)
5. Gollucci, P.M.: Apache.org incident report for 04/09/2010. https://blogs.apache.org/infra /entry/apache_org_04_09_2010 (2010)
6. Hirsch, F.: Device APIs Working Group. http://www.w3.org/2009/dap/ (2014)
7. Infosecurity: Adobe hacked customers' card details and adobe source code stolen. http://www.infosecurity-magazine.com/view/34872/adobe-hacked-customers-card-details-and-adobe-source-code-stolen (2013)
8. Infosecurity: How GCHQ hacked belgacom. http://www.infosecurity-magazine.com/view/355 58/how-gchq-hacked-belgacom (2013)
9. Jacobs, F.: How reuters got compromised by the syrian electronic army. https://medium.com/ @FredericJacobs/the-reuters-compromise-by-the-syrian-electronic-army-6bf570e1a85b(2014)
10. Kelly, S.M.: LastPass passwords exposed for some internet explorer users. http://mashable.com /2013/08/19/lastpass-password-bug/ (2013)
11. King, A.: Club nintendo japan hacked, user details could be compromised. http://wiiudaily.com /2013/07/club-nintendo-japan-hacked/ (2013)
12. Kovacs, E.: Vodafonegermany hacked, details of 2 million users stolen. http://news.softpedia. com/news/Vodafone-Germany-Hacked-Details-of-2-Million-Users-Stolen-382458.shtml (2013)
13. Luo, T., Hao, H., Du, W., Wang, Y., Yin, H.: Attacks on webview in the android system. In: Proceedings of the 27th Annual Computer Security Applications Conference (ACSAC), pp. 343–352 (2011)
14. Martin, B., Brown, M., Paller, A., Kirby, D.: Cwe/sans top 25 most dangerous programming errors. http://cwe.mitre.org/top25/ (2011)
15. Masnick, M.: FLYING PIG: The NSA is running man in the middle attacks imitating Google's servers. http://www.techdirt.com/articles/20130910/10470024468/flying-pig-nsa-is-running-man-middle-attacks-imitating-googles-servers.shtml (2013)
16. Perlroth, N.: Lax security at linkedin is laid bare. http://www.nytimes.com/2012/06/11/technol ogy/linkedin-breach-exposes-light-security-even-at-data-companies.html?pagewanted=all (2012)
17. Reisinger, D.: eBay hacked, requests all users change passwords. http://www.cnet.com/news/ ebay-hacked-requests-all-users-change-passwords/ (2014)
18. Riley, M., Elgin, B., Lawrence, D., Matlack, C.: Missed alarms and 40 million stolen credit card numbers: How target blew it. http://www.businessweek.com/articles/2014-03-13/target-missed-alarms-in-epic-hack-of-credit-card-data (2014)
19. Ross, D.: Entry point regulation for web apps. http://randomdross.blogspot.be/2014/08/entry-point-regulation-for-web-apps.html (2014)
20. Sterne, B., Barth, A.: Content Security Policy 1.0. W3C Candidate Recommendation (2012)
21. Symantec Corporation: 2013 norton report. http://www.symantec.com/about/news/resources/ press_kits/detail.jsp?pkid=norton-report-2013 (2013)
22. Symantec Corporation: Internet security threat report. http://www.symantec.com/content/en/ us/enterprise/other_resources/b-istr_main_report_v19_21291018.en-us.pdf (2014)
23. TNS Opinion & Social: Special eurobarometer 404—cyber security. http://ec.europa.eu/ public_opinion/archives/ebs/ebs_404_en.pdf (2013)
24. Wichers, D.: Owasp top 10. https://www.owasp.org/index.php/Category:OWASP_Top_Ten_ Project (2013)

Chapter 2
Traditional Building Blocks of the Web

The previous chapter introduced the Web, covering the evolution from static Web pages towards a dynamic application platform. This chapter is more technical and provides the necessary background on how the Web works, which will help you in understanding the nuances of the attacks covered in the later chapters.

Within the distributed, hypertext-based Web, we will focus on the client-side features that enabled the Web to evolve into the dynamic application platform it is today and the browser security policies that are supposed to keep Web applications in line. Many of the topics covered in this chapter have been introduced in the early stages of the Web's development but are still present or reused in modern Web applications.

This chapter will first cover the basic building blocks of traditional Web applications, offering details on how content is loaded, how users can be authenticated and how session management mechanisms enhance the stateless HTTP protocol. Next, we will cover the browser's security policies, which regulate what Web applications can do within the browser, up to this day. We also investigate how client-side features can be extended beyond HTML, both by plugins for arbitrary content and by browser extensions. Finally, we cover several browser features that enhance the user's window on the Web.

2.1 Traditional Web Technology

Most modern Web applications are highly dynamic, process content in the background and fetch information on a continuous basis. While these applications seem vastly different from traditional Web applications, they share the same basis, and still use the same underlying concepts. This section briefly explains these traditional building blocks, offering you the required background knowledge.

© Philippe De Ryck, Lieven Desmet, Frank Piessens, Martin Johns 2014
P. De Ryck et al., *Primer on Client-Side Web Security*,
SpringerBriefs in Computer Science, DOI 10.1007/978-3-319-12226-7_2

2.1.1 Loading Web Content

Content on the Web is identified by a *Uniform Resource Identifier* (URI), a more general form of the earlier *Uniform Resource Locators* (URL). An example of a URI is *http://example.com/thisbook.html*. The first part of the URI, before the *://* is called the *scheme* and identifies the protocol to be used for fetching the resource. Most URIs on the Web today use the *http* or *https* scheme, which identifies HTTP protocol [18], either over a plaintext channel or over a secure channel, using Transport Layer Security [15], a topic that will be discussed in more detail in Chap. 5. Whenever the browser wants to load such a resource, it issues an HTTP request to the remote server, which is identified by the next part of the URI, between the *://* and the next */*, here *example.com*. The server at this address responds with an appropriate HTTP response for the requested path and parameters, which are the last part of the URI, here *thisbook.html*. This request/response-based communication protocol lies at the basis of every communication on the Web, even today.

HTTP requests and responses follow a certain pattern but have many configurable fields. A request has a certain method, such as GET, to retrieve information and POST to submit data to the server. In addition, both requests and responses can have headers, carrying meta-information about the request. We will not discuss all these possibilities in detail, information that can easily be found in other reference works [36]. One specific characteristic of HTTP that is relevant for the remainder of this text is that the protocol is *stateless*, meaning that there is no relation or required order between subsequent requests. Any need for relations or order between requests needs to be maintained by the browser and/or servers, on top of HTTP, for example by using request and response headers.

HTTP/2.0 [8], currently under development by the IETF, will bring several significant improvements to HTTP protocol, while preserving the original protocol's semantics. HTTP/2.0, based on the SPDY protocol by Google [7], essentially changes the way HTTP traffic is sent on the wire, reducing the page load time. HTTP/2.0 introduces new features such as multiplexed requests, prioritized requests, compressed headers, and support for server-pushed content. On the security side, a secure channel using TLS has not been made mandatory, but the most efficient upgrade path from the current HTTP/1.1 to HTTP/2.0 is by deploying HTTP/2.0 over TLS using the *Application Layer Protocol Negotiation* extension [19].

2.1.2 Authentication and Authorization

Even in the early Web, when sites offered only static content, authentication and authorization could be used to restrict access to the provided content. HTTP protocol provides the *Authorization* request header, aimed at providing the Web application with the user's credentials. The most common authentication scheme with the *Authorization* header uses *Basic* authentication, where the username and password are base64-encoded,[1] and included as the header value. This allows an application to extract these credentials, verify them, and make a decision on whether to allow the request or not. After a successful authentication, the browser will attach the user's credentials to every subsequent request to this origin. Since the browser remembers these credentials during its session, logging out of an account is only possible by closing the browser.

As Web applications became more complex, developers wanted to integrate authentication with the application, streamlining the user experience within the same look and feel. To authenticate users, they embedded an HTML form, where the user had to enter a username and a password. By submitting the form, the username and password were sent to the server, where they could be validated. However, since the credentials are only sent in a single request after form submission, instead of in every subsequent request as with the *Authorization* header, Web applications needed a way to remember the user's authentication state. Keeping track of the authentication state is solved by using session management, our next topic of discussion.

2.1.3   Cookies and Session Management

As mentioned before, subsequent requests in HTTP protocol are independent of each other. The lack of any relation between requests and the incapability of keeping state between requests has resulted in the introduction of cookies [4]. Cookies are server-provided key-value pairs, stored by the client, attached to every request to the same domain. Essentially, cookies allow the server to store some state at the client side, which will be attached to future requests. For example, cookies can be used to store a language preference, allowing the Web application to serve the requested content in the desired language.

A more complex mechanism, nowadays built on top of cookies, is session management. A session management mechanism offers a server-side session object, and associates multiple requests from the same user with this server-side session object. Web applications can use the session object to store useful session information, such as an authentication state, a shopping cart, etc.

Under the hood, session management mechanisms assign a random, unique session identifier to a newly created session. The session identifier is sent to the client in a cookie and will be attached to every subsequent request. By looking up the session object that belongs to the session identifier in the request, the Web application can process the request in the appropriate context.

Session management mechanisms based on session identifiers were already available before cookies were widely supported. These mechanisms included the session

---

[1] Base64 encoding transforms the entered username and password into an alphanumeric string, which is easily reversed. The credentials are not encrypted, as is often mistakenly believed.

identifier as a parameter in the URI, where it could be extracted by the Web application. These mechanisms have seen a slow demise because of practicality reasons, since every URI in the Web application needed to be dynamically generated to include the user-specific session identifier. In addition, embedding the identifier in the URI also holds a security risk, since the URI is easily leaked or copy/pasted. Note that many session management mechanisms still offer parameter-based session management as a fallback mechanism in case a browser does not support cookies.

2.2   Browser Security Policies

Modern browsers are the execution platform for complex Web applications, which consist of different kinds of static and dynamic content, coming from multiple providers with varying trust levels. Within the browser, several security policies govern the behavior of this content, regulating interactions between different contexts, managing access to potentially sensitive resources, and preventing unauthorized navigation attempts. These security policies are essential for client-side Web security, as their subtle nuances are often abused in attacks, and their restrictions are relied upon when building countermeasures.

Browser security policies generally depend on the notion of an *origin*. An origin is defined as the triple *(scheme, host, port)*, which are part of any URI,[2] in this section, we cover the three most important browser security policies. The *Same-Origin Policy* prevents unrestricted interactions between contexts from different origins, and regulates access to sensitive resources and application programming interfaces (APIs) on the basis of the origin of a document. This is the core security policy of the browser. Second, we discuss how the browser deals with the inclusion of cross-origin content, a common practice in almost every modern Web application. Third, we cover the *context navigation policy*, which is responsible for preventing unauthorized navigation requests between nested contexts.

2.2.1   Same-Origin Policy

The core security policy in a browser is the Same-Origin Policy (SOP), which regulates direct interactions between different browsing contexts. The basic function of the SOP is to prevent scripts loaded in one origin from programmatically accessing resources from other origins. For example, if you have a script running on a page

---

[2] The port is an optional URI component, and when omitted, the protocol's default port is used, which is 80 for HTTP and 443 for HTTPS.

loaded from the URI *http://www.example.com*, it is not allowed to access the resources of a page loaded from *http://www.secret.com*, as the origins of both contexts are different, due to the distinct hosts.

There is one way to relax the constraints of the SOP, by using the *document.domain* JavaScript property, which allows two Web applications that share the same parent domain to interact with each other. For example, the application at *www.example.com* and *login.example.com* can both set their *document.domain* property to *example.com*, overriding any future same-origin checks with this parent domain. This allows two sibling applications to cooperate freely. Even though both parties must explicitly opt-in to this feature, once they have opted-in, any other site within the same parent domain can "join" as well.

The SOP originally started as a way to prevent access to the document object model (DOM) but has been gradually extended to other resources accessible within the origin. One such example is the recently introduced *canvas* element in HTML 5 [9], which allows the Web application to use JavaScript code to draw graphics and extract the result as an image. Certain features of the canvas allow the script to draw arbitrary, cross-origin resources on the canvas (e.g., an image or video), making it possible to steal the contents of the video or image. To prevent such cross-origin leaking, the specification requires the browser to consider the canvas to be "tainted" with the origin of the image, effectively preventing any access from an origin other than that of the image.

Another example is the XMLHttpRequest (XHR) object [35], which allows JavaScript code to issue new HTTP requests. Traditional XHR requests can only be sent to servers within the same origin as the origin of the document containing the script, motivated by the high degree of flexibility offered by the XHR object. Failing to restrict these requests would allow a malicious page to send custom HTTP requests to unsuspecting servers, for example, using the PUT or DELETE methods. As we will discuss in more detail in the next chapter, these restrictions have recently been relaxed with a server-driven security policy [34], inadvertently causing problems for applications that implicitly depended on this same-origin restriction [3].

Finally, the security policy for cookies is similar to origin-based policies but is actually domain-based. Cookies are generally set for a domain and only sent to the corresponding domain. A similar situation for script-based cookie access exists: any application that resides on the domain of the cookies, or a valid subdomain, can access the cookies from JavaScript. For example, cookies set explicitly for *www.example.com* can not only be accessed by any resource in *www.example.com* but also by resources under *dev.www.example.com*.

2.2.2 Security Model for Third-Party Content Inclusion

Modern Web applications include content from a wide variety of locations, often residing within a different origin. Common examples of such third-party content inclusions are images, style sheets, or JavaScript files, simply integrated by including an HTML tag with the appropriate URI. The inclusion of various JavaScript libraries is especially popular, since it allows the creation of highly responsive user interfaces, and enriches the Web site with additional functionality, ranging from integration with social media sites, to context-sensitive advertisements and tools for Web site analytics.

There are two commonly-used techniques to integrate third-party JavaScript into a Web application: through *script inclusion* or via *iframe integration*. The former loads the script within the security context of the including application, resulting in a straightforward way to integrate components and enable interaction between components. The latter places the script in a separate frame, within its own security context, effectively shielding sensitive resources, but making interaction a bit more complicated. We elaborate on both techniques below.

Script Inclusion

HTML *script* tags are used to include and execute JavaScript while a Web page is loading. This JavaScript code can be located on a server with a different origin than the integrating page. When executing, the browser will treat the code as if it originated from the same origin as the Web page itself, without any restrictions of the SOP.

The included code executes in the same JavaScript context, and has access to the code of the integrating Web page and all of its data structures. All sensitive JavaScript operations available to the integrating Web page are also available to the integrated code.

Prevalence of Third-Party Script Inclusion

A 2012 study [25] examined 3,300,000 pages of the top 10,000 Alexa sites, and analyzed 8,439,799 remote script inclusions. From the results (shown in Fig. 2.1), it becomes clear that 88.45 % of the 10,000 Web sites included at least one remote JavaScript library. Even more remarkable, some sites in the top Alexa list trust up to 295 unique remote hosts.

The caveat that applies to third-party content inclusion is the interaction with the security policies of the browser, mainly the SOP. The included content generally resides directly within the security context of the including document, which is not a problem for static content, but results in complications when dynamic content, such as JavaScript, is included. The dynamic code is included within the security context of the application, where it gains access to all origin-restrained resources. Combined with the practice of including numerous third-party JavaScript libraries, this is a major security challenge for the Web.

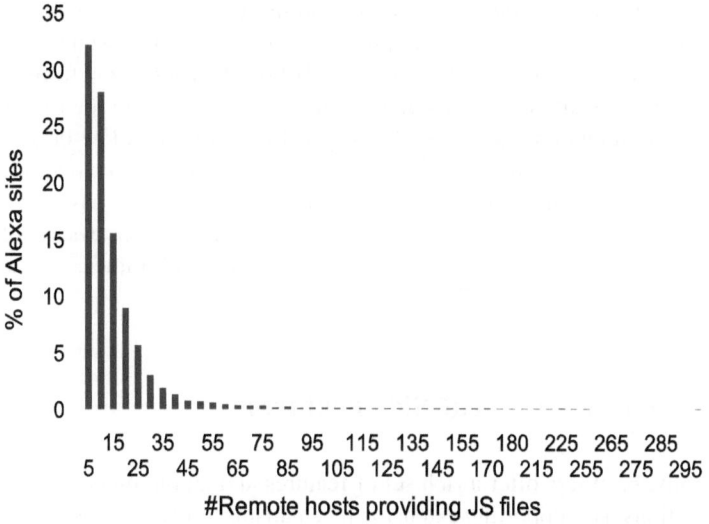

Fig. 2.1 Relative frequency distribution of the percentage of top Alexa sites and the number of unique remote hosts from which they request JavaScript code. (Figure copyrighted by ACM, published in [25] with DOI *10.1145/2382196.2382274*)

Iframe Integration

HTML *iframe* tags allow a Web developer to include one document inside another. The integrated document is loaded in its own environment almost as if it were loaded in a separate browser window. The advantage of using an *iframe* in a Web application is that the integrated component (coming from another origin) is isolated from the integrating Web page by the SOP. However, the code running inside the *iframe* still has access to the available JavaScript APIs, albeit limited within its own execution context (i.e., origin). For instance, a third-party component can use local storage APIs but has access only to the local storage of its own origin and not to those of the integrating page.

The newly introduced HTML 5 *sandbox* attribute [10] aims to support the embedding of untrusted content in an *iframe* by putting security restrictions on the *iframe*, such as disabling JavaScript, turning off plugins, and restricting navigation. Through coarse-grained directives, several features can be re-enabled, for example by specifying the "allow-scripts" keyword to enable JavaScript.

2.2.3   Context Navigation Policy

Navigation events occur frequently on the Web, for example when a user opens a page, follows a link, or when an automatic redirect happens. Triggering navigation events from within a document's context is straightforward, for example using

JavaScript to modify the *document.location* property or by automatically following a link. Navigation becomes more complicated when one context wants to navigate a window or frame from another context, possibly hosting a document from a different origin. Common examples are documents that want to navigate their child frames or a popup window they own. The decision as to whether to allow or deny such a navigation is not based on the SOP, which would prohibit any navigation between contexts from different origins, but is determined by a separate navigation policy. Several navigation policies have been proposed, but modern browsers all use the *descendant policy* [6, 9], which restricts navigation to child frames, or frames with an equivalent level of access [5].

2.3   Extending the Client-Side Features

Traditionally, browsers offer a rich set of features and ample functionality but also have limitations. By supporting extension mechanisms, browsers give developers the ability to enhance the browser experience, adding additional features. A first popular way of extending the browser is by adding plugins to handle arbitrary content. The most popular example of a browser plugin is Adobe's Flash player, which is capable of playing Flash files, which add dynamic content to a Web page. The second extension mechanism is browser extensions, which allow the user to modify the core behavior of the browser, adding additional features, or user interface (UI) items. Popular examples of browser extensions are NoScript, to limit the JavaScript that is run on a Web page, and AdBlock, to prevent intrusive advertisements from being loaded.

While these mechanisms clearly extend the functionality of the browser, they also have their consequences. For instance, they significantly enlarge the attack surface, as demonstrated by regular discoveries of malicious or vulnerable plugins or extensions [20, 31–33]. In this section, we briefly discuss how plugins and extensions work, how they are integrated in the browser and what consequences are associated with their use.

15	**Embedding Advertisements in a Web Application**
16	Many of the free Web applications are built on an advertisement-based business
17	model, where third-party advertisements are embedded in the pages of the
18	application. These advertisements are created by the companies that want to
19	advertise their services or products, and are delivered through advertisement
20	networks, such as DoubleClick, AdSense, and AdBrite.

Unfortunately, the Web application embedding the advertisements has no control of the content, and hence embeds untrusted content into its pages. Integrating untrusted content in a Web application comes with a trade-off between flexibility and security [13]. Using script inclusion for integrating the ads gives the advertisement provider great flexibility in ad placement, as well as the capability to offer content-specific advertisements based on the displayed content of the embedding page. *iframe* integration offers the embedding page the necessary security guarantees but puts certain restrictions on the embedded content. The rigidity of *iframes* and the lack of security guarantees of scripts have driven the research community to come up with alternative approaches that offer isolation guarantees but enable the interaction demanded by advertisement networks [2, 16, 29]. From a high-level point of view, these approaches isolate the advertisement from the main page using a sandbox technique and allow a filtered set of interactions with the page's content. The most important interactions are drawing the advertisement in a dedicated part of the page and receiving the user's interaction with the advertisement, for example, clicking on the advertisement.

While these research efforts provide viable alternatives to the traditional *script* and *iframe*-based integration techniques, they are not used in practice. Almost every advertisement network uses *script*-based advertisements; this which occasionally results in the spreading of a malicious advertisement [21, 23, 27].

2.3.1 Plugins for Arbitrary Content

Browser plugins are generally designated handlers for specific kinds of content. The most common example of a browser plugin is Adobe's Flash Player, responsible for processing and displaying Flash content within the browser. Other popular examples are Silverlight, Java, ActiveX, and PDF reader plugins. Browser plugins are associated with MIME types and are automatically invoked when the content of a specified MIME type is encountered. The content is subsequently processed in the plugin's own runtime environment. A browser plugin can provide arbitrary functionality and is not limited to content rendering. An example of a non-content rendering plugin is the *Gnome Shell Integration* plugin [30], supporting the installation of additional widgets from the distribution site into the Gnome desktop environment.

Plugin content is embedded in a document, and the registered handler is triggered by the browser when this content is encountered. Most plugins allow communication between the document and the plugin using JavaScript, albeit with some restrictions, depending on the implementation. For example, in the case of Flash, an interface can be exposed towards the document, and arbitrary JavaScript functions can be executed in the embedding page.

Support for browser plugins is widespread on computing platforms running a traditional operating system, such as notebooks and desktop machines. Mobile support for browser plugins is extremely limited. For example, Apple's iOS does not support any browser plugins, and Adobe has abandoned efforts for supporting Flash on Android 4.1 and higher [11]. The demise of Flash on mobile platforms can mainly be attributed to the rise of HTML 5's dynamic content features, and the performance issues associated with running a Flash player on a mobile device. Microsoft's Windows Phone only supports the in-house Silverlight plugin. Due to the limited support of plugins on mobile devices, Google has started to optimize search results, indicating which pages are likely to cause problems for your mobile device [37].

Plugin content runs within the environment of the handler, where the security policies of the browser no longer reign. This effectively means that if the plugin does not restrict the behavior of plugin content, basic browser policies are easily circumvented. One example is Flash, which allows developers to specify a policy to enable cross-origin requests, regardless of the SOP restrictions. A server can define the *crossdomain.xml* policy file, defining the origins from where remote requests are accepted. The Flash plugin is responsible for checking the file before carrying out the cross-origin request.

Plugins can also be a source of severe security problems, as illustrated by the numerous Java vulnerabilities in 2013 [31], eventually even leading to browser vendors recommending that Java should be disabled altogether. One potential source of vulnerabilities is the inability to deal with untrusted and potentially malicious input [36]. Therefore, close cooperation between browser vendors and plugin developers is crucial. In recent developments, the security of the Flash plugin has been significantly tightened. Flash is now effectively sandboxed on the OS level, preventing serious harm in case a vulnerability is found and exploited [22]. Alternatively, a new trend is emerging whereby plugin content is initially automatically disabled, but the user is then given the option to activate each piece of content separately, by a single click on a displayed *play* button [12].

2.3.2 Browser Extensions

Browser extensions extend the core functionality of the browser and come in various flavors, from a simple toolbar to behavior-changing extensions. An extension is not associated with a MIME content type but uses the exposed APIs to register hooks and react to events. Some popular examples of extensions are NoScript, which selectively disables JavaScript on Web pages; AdBlock, which removes advertisements; or FireBug, a Web development tool offering a debugger, giving a view on network traffic, etc.

Extensions for the Chrome and Firefox browsers are written in JavaScript,[3] and are restricted by the API offered by the browser. On Firefox, extensions can access almost all browser internals and can also access the file system or launch commands on the operating system. Chrome follows a more conservative approach, offering

---

[3] Native code is also supported but discouraged since it requires different versions for different platforms.

access to a select number of browser events but preventing extensions from reaching outside the browser.

Browser extensions are very powerful; first of all because they can potentially access everything that happens within the browser. In addition, Firefox extensions can also access other resources on the user's machine, making them even more powerful. Preventing every form of misuse is virtually impossible, even with the manual verification system employed by the Mozilla Add-Ons Web site. Therefore, installing a browser extension effectively enlarges the attack surface, a risk unknown to or accepted by the users. One example is extensions in combination with private browsing mode, which enable users to browse without leaving a trace on the local machine. Unfortunately, many extensions fail to correctly deal with private browsing mode, potentially exposing private information once the session is terminated [24]. Due to such potentially unexpected violations, Google Chrome automatically disables extensions in private browsing mode, giving users the option to explicitly enable them, if desired.

Contrary to plugins, which are often cross-browser runtime environments, browser extensions are browser-specific. Browser support for extensions is less widespread than support for plugins, but extensions are better supported on mobile browsers. On the traditional platforms, Chrome, Firefox, and Opera offer the most extensive support for extensions, along with rich APIs. On the mobile platforms, Firefox and Opera offer extensive support for extensions through a similar mechanism as on traditional operating systems. They typically do require a separate user interface, adapted to the screen and interaction patterns of a mobile device. The mobile version of Chrome does not support extensions, and there are no plans to change this at the time of writing.

2.4 Enhancing the User's Window on the Web

Users have a straightforward way to access and interact with Web content through the browser. Next to the underlying technicalities of actually fetching, rendering, and securing the content, the browser also offers several user-targeted features, aimed at improving the browsing experience. In this section, we cover an important security feature, the secure sockets layer (SSL)/transport layer security (TLS) indicator, which attempts to warn the user of suspicious or unsafe activities. We also discuss private browsing modes, where a user can browse the Web without leaving a local trace. The last feature we briefly cover in this section is the synchronization of features across browsers running on multiple devices.

SSL/TLS Security Indicators

Traditionally, a browser displayed a small lock icon to indicate that a Web site was loaded over a SSL/TLS-secured connection. As the features of certificates changed, for example, with extended validation certificates, the security indicators slightly

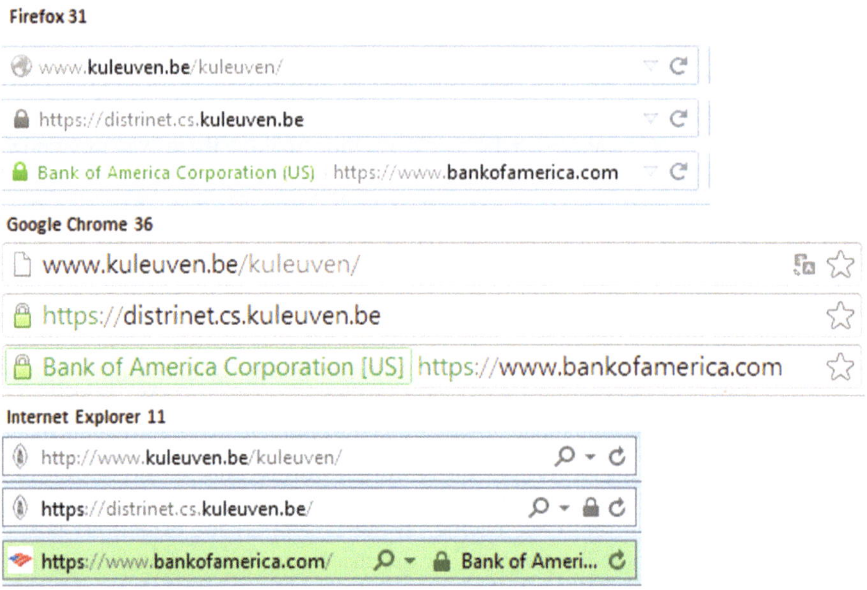

Firefox 31

Google Chrome 36

Internet Explorer 11

Fig. 2.2 Browser vendors choose different ways of indicating the level of trust in the SSL/TLS certificate, making it hard for users to grasp the precise meaning

changed as well. Currently, browsers use a combination of colors, lock icons, and company names to indicate the level of security (see Fig. 2.2). Unfortunately, these security indicators are often confusing and misunderstood [14, 28], therefore missing the intended effect. This topic has once again become highly relevant with the rise of browsers on mobile devices with limited screen sizes.

Private Browsing Modes

When opening a window in *private browsing mode*, also known as *incognito mode* or *InPrivate*, the browser creates a window that allows the user to browse the Web, without leaving a local trace after terminating the session. For example, all newly created cookies are removed, and no entries appear in the browser's history. The implementation detail of the information that is available at the beginning differs per browser, but the general concept remains the same. Note that private browsing mode is aimed at offering privacy on the local machine, but does not necessarily protect the user's identity towards the server. While the cookies from normal browsing sessions may not be available in private browsing mode, alternative techniques are available to keep track of users. One example is browser fingerprinting, which uses a set of browser characteristics to compile a "fingerprint" [1, 17, 26].

Cross-Browser Synchronization

Recently, browsers have started supporting synchronization services, allowing users to share bookmarks, stored credentials, and open tabs across multiple installations of a browser. Chrome takes this feature one step further, by also sharing devices connected to the machine of a running browser instance. For example, leaving a Chrome instance running on your workstation at work, allows you to print at work from your Chrome instance at home.

References

1. Acar, G., Juarez, M., Nikiforakis, N., Diaz, C., Gürses, S., Piessens, F., Preneel, B.: Fpde-tective: dusting the web for fingerprinters. In: Proceedings of the 20th ACM Conference on Computer and Communications Security (CCS), pp. 1129–1140 (2013)
2. Agten, P., Van Acker, S., Brondsema, Y., Phung, P.H., Desmet, L., Piessens, F.: JSand: complete client-side sandboxing of third-party JavaScript without browser modifications. In: Proceedings of the 28th Annual Computer Security Applications Conference (ACSAC), pp. 1–10 (2012)
3. Austin, M.: Hacking facebook with HTML5. http://m-austin.com/blog/?p=19 (2010)
4. Barth, A.: HTTP state management mechanism. RFC Proposed Standard (RFC 6256) (2011)
5. Barth, A., Jackson, C.: Protecting browsers from frame hijacking attacks. http://seclab.stanford.edu/websec/frames/navigation/ (2008)
6. Barth, A., Jackson, C., Mitchell, J.C.: Securing frame communication in browsers. Commun. ACM **52**(6), 83–91 (2009)
7. Belshe, M., Peon, R.: SPDY protocol. IETF Internet Draft (2012)
8. Belshe, M., Thomson, M., Melnikov, A., Peon, R.: Hypertext transfer protocol version 2.0. IETF Internet Draft (2014)
9. Berjon, R., Faulkner, S., Leithead, T., Navara, E.D., O'Connor, E., Pfeiffer, S., Hickson, I.: HTML 5.1 specification. W3C Working Draft (2014)
10. Berjon, R., Faulkner, S., Leithead, T., Navara, E.D., O'Connor, E., Pfeiffer, S., Hickson, I.: HTML 5.1 specification—the sandbox attribute. W3C Working Draft (2014)
11. Brewis, M.: How to add adobe flash to an android phone or tablet. http://www.pcadvisor.co.uk/how-to/google-android/3417930/flash-on-android/ (2014)
12. Coates, M.: Putting users in control of plugins. https://blog.mozilla.org/security/2013/01/29/putting-users-in-control-of-plugins/ (2013)
13. De Ryck, P., Decat, M., Desmet, L., Piessens, F., Joosen, W.: Security of web mashups: A survey. In: Proceedings of the 15th Nordic Conference on Secure IT Systems (NordSec), pp. 223–238 (2010)
14. Dhamija, R., Tygar, J.D., Hearst, M.: Why phishing works. In: Proceedings of the ACM CHI conference on human factors in computing systems (CHI), pp. 581–590 (2006)
15. Dierks, T., Rescorla, E.: The transport layer security (TLS) protocol version 1.3. RFC 5246bis (2014)
16. Dong, X., Tran, M., Liang, Z., Jiang, X.: Adsentry: Comprehensive and flexible confinement of javascript-based advertisements. In: Proceedings of the 27th Annual Computer Security Applications Conference (ACSAC), pp. 297–306 (2011)
17. Eckersley, P.: How unique is your web browser? In: Proceedings of the 10th Privacy Enhancing Technologies Symposium (PETS), pp. 1–18 (2010)
18. Fielding, R., Gettys, J., Mogul, J., Frystyk, H., Masinter, L., Leach, P., Berners-Lee, T.: Hypertext transfer protocol—HTTP/1.1. RFC 2616 (1999)

19. Friedl, S., Popov, A.: Transport Layer Security (TLS) application layer protocol negotiation extension. RFC Proposed Standard (RFC 7301) (2014)
20. Heath, N.: Malicious Chrome and Firefox extensions found hijacking Facebook profiles. http://www.zdnet.com/malicious-chrome-and-firefox-extensions-found-hijacking-facebook-profiles-7000015277/ (2013)
21. Jacobs, F.: How reuters got compromised by the syrian electronic army. https://medium.com/@FredericJacobs/the-reuters-compromise-by-the-syrian-electronic-army-6bf570e1a85b (2014)
22. Keizer, G.: Google builds stronger Flash sandbox in Chrome. http://www.computerworld.com/s/article/9230094/Google_builds_stronger_Flash_sandbox_in_Chrome (2012)
23. Kirk, J.: Yahoo's malware-pushing ads linked to larger malware scheme. http://www.pcworld.com/article/2086700/yahoo-malvertising-attack-linked-to-larger-malware-scheme.html (2014)
24. Lerner, B., Elberty, L., Poole, N., Krishnamurthi, S.: Verifying Web browser extensions compliance with private-browsing mode. In: Proceedings of the 18th European Symposium on Research in Computer Security (ESORICS), pp. 57–74 (2013)
25. Nikiforakis, N., Invernizzi, L., Kapravelos, A., Van Acker, S., Joosen, W., Kruegel, C., Piessens, F., Vigna, G.: You are what you include: large-scale evaluation of remote Javascript inclusions. In: Proceedings of the 19th ACM conference on Computer and communications security, pp. 736–747 (2012)
26. Nikiforakis, N., Kapravelos, A., Joosen, W., Kruegel, C., Piessens, F., Vigna, G.: Cookieless monster: Exploring the ecosystem of web-based device fingerprinting. In: Proceedings of the 34th IEEE Symposium on Security and Privacy (SP) (2013)
27. Rubenking, N.: Black hat briefing: building a million browser botnet for cheap. http://securitywatch.pcmag.com/security/314341-black-hat-briefing-building-a-million-browser-botnet-for-cheap (2013)
28. Schultze, S.: Web browser security user interfaces: Hard to get right and increasingly inconsistent. https://freedom-to-tinker.com/blog/sjs/web-browser-security-user-interfaces-hard-get-right-and-increasingly-inconsistent/ (2011)
29. Ter Louw, M., Ganesh, K.T., Venkatakrishnan, V.: AdJail: Practical Enforcement of Confidentiality and Integrity Policies on Web Advertisements. In: Proceedings of the 19th USENIX Security Symposium, pp. 371–388 (2010)
30. The GNOME Project: What's this?—GNOME shell extensions. https://extensions.gnome.org/about/ (2013)
31. US-CERT: Oracle Java Contains Multiple Vulnerabilities. Alert (TA13-064A) (2013)
32. Van Acker, S., Nikiforakis, N., Desmet, L., Joosen, W., Piessens, F.: Flashover: automated discovery of cross-site scripting vulnerabilities in rich internet applications. In: Proceedings of the 7th ACM Symposium on Information, Computer and Communications Security (ASIACCS), pp. 12–13. ACM (2012)
33. Van Acker, S., Nikiforakis, N., Desmet, L., Piessens, F., Joosen, W.: Monkey-in-the-browser: malware and vulnerabilities in augmented browsing script markets. In: Proceedings of the 9th ACM symposium on Information, computer and communications security (ASIACCS), pp. 525–530. ACM (2014)
34. van Kesteren, A.: Cross-origin resource sharing. W3C Recommendation (2014)
35. van Kesteren, A., Aubourg, J., Song, J., Steen, H.R.M.: XMLHttpRequest. W3C Working Draft (2014)
36. Zalewski, M.: The Tangled Web: A Guide to Securing Modern Web Applications. San Francisco, No Starch Press (2012)
37. Zeckman, A.: New Google mobile alert: Websites using flash may not work on your device. http://searchenginewatch.com/article/2355766/New-Google-Mobile-Alert-Websites-Using-Flash-May-Not-Work-on-Your-Device (2014)

Chapter 3
The Browser as a Platform

In the previous chapters, we saw how the Web has evolved into a dynamic platform governed by browser security policies. This evolution is enabled by the browser which gives the user access to the Web. Dominating in the browser market has been a high-stakes game for major browser vendors. In the first decade of the twenty-first century, the *browser wars* were at an all-time high, resulting in incompatible implementations of Web standards. Fortunately, browser vendors have become more cooperative with the development of the HTML 5 specification [2], leading to a widely supported specification, adding numerous new features.

In this chapter, we look at the browser from various perspectives. We discuss how the browser is deployed on numerous devices nowadays, bringing the Web everywhere. We explain how the browser grew into a feature-rich application platform, offering features such as client-side storage, remote communication mechanisms, mobile features, and even the registration of default applications for handling certain types of content. Finally, we take a look at the latest evolution, where the browser became similar to an operating system, with Firefox OS and Google's Chrome OS being two major examples.

3.1 The Synergy Between Browsers and Devices

The browser has become one of the most important applications on a computer, but has also conquered other devices. Nowadays, browsers can be found on numerous devices, such as mobile phones, tablets, gaming consoles, smart televisions, and even in modern cars, with many more to come. Some of these devices are open platforms and support multiple browsers, while others are closed and only support a single browser, or even provide their own custom software. Having a browser on all these devices not only creates easy access to the Web but also brings certain security risks. Browsers are known to have vulnerabilities, which can be exploited by malicious Web pages. As a consequence, browsers require frequent updates, well-established on traditional platforms, but more challenging on some modern platforms, such as cars or smart TVs.

© Philippe De Ryck, Lieven Desmet, Frank Piessens, Martin Johns 2014
P. De Ryck et al., *Primer on Client-Side Web Security,*
SpringerBriefs in Computer Science, DOI 10.1007/978-3-319-12226-7_3

All the major browser vendors offer different flavors of their flagship browser, tailored to the specific devices they support. For example, Google has a mobile version of Chrome for mobile devices, with some limitations, such as no support for extensions, while Firefox offers the Fennec browser, a mobile version of Firefox, with roughly the same feature set, but a different UI. Similarly, household devices embedding a browser generally cooperate with a browser vendor, allowing for product-specific customs when necessary.

When digging deeper into the different browser vendors and their variety of browser products, we end up at the heart of the browser, the *browser engine* or *layout engine*. The browser engine is responsible for actually processing the Web content, resulting in a rendered version of the page, and the execution of any embedded scripts. The four most well-known browser engines are Gecko (Firefox), Trident (Internet Explorer), WebKit (Safari), and Blink (Chrome and Opera), which was recently forked from WebKit. In an ideal world, all these browser engines would implement the same set of features defined by the W3C standardization committee. In reality, they support a large subset of all these specifications, have their specific quirks with certain features, and implement some vendor-specific features [25], making cross-browser Web development challenging.

A recent trend observed not only in mobile devices but also in traditional desktop applications is the integration of a browser-rendering engine into a custom-developed application. The rendering engine is offered as a library that can be included and can then be instructed to load Web content. This is often used for displaying advertisements or documentation pages in a native application. However, since the introduction of the powerful HTML 5 specification [2], app developers have started writing the bulk of their code using HTML, CSS, and JavaScript, and render this code using a small native app, which simply loads the rendering library. This native application is able to expose certain operating system APIs to the embedded browser engine, offering the JavaScript code access to storage, contacts, cell phone operations, etc. Further evolution of this powerful development model has led to cross-device frameworks for Web applications, such as Apache Cordova and Phone-Gap. These frameworks allow a developer to wrap an app built with Web technology into native applications for numerous mobile platforms, such as iOS, Android, Windows Phone, BlackBerry, etc.

In the end, all browsers are built on the same concepts of the Web, regardless of the device they run on or the environment they are deployed in. Every browser relies on the same cornerstone security policies to govern the Web application it loads. Even the state-of-the-art features, being introduced today and tomorrow, base their security guarantees on concepts such as the Same-Origin Policy and origin-based access control.

3.2 From Rendering Engine to Feature-Rich Platform

Traditional browsers focused on fetching content and rendering it to the screen. Embedded JavaScript would be executed, but it was limited to manipulating the rendered page through the DOM, fetching additional content and accessing the cookies. With the evolution of the Web, the available functionality at the client side has evolved as well, offering advanced APIs and system features to the script environments. This results in a full-featured client-side platform, where Web applications become as powerful as traditional native applications. Take for example Google's productivity suite which contains a text editor, spreadsheet, and presentation software [9].

In this section, we cover the available features in several important areas, such as storage, communication, mobile features, and default content handlers. Besides the specifications we discuss here, there are many more, such as background processing [12], sending notifications through the OS's notification mechanism [22], and performing cryptographic operations [6].

3.2.1 Client-Side Storage

Traditionally, Web applications did not have access to persistent storage or caching mechanisms on the client. Some Web applications used cookies to store client-side data, but the disadvantage is that this data is sent to the server on every request, causing an enormous overhead. In response to the need for client-side storage, several storage mechanisms have been made available, including an application cache for offline applications.

The Web Storage specification [13] offers a straightforward key/value-pair storage mechanism. Web Storage offers both persistent storage using the *localstorage* object and temporary storage using the *sessionstorage* object. While Web Storage is supported on most browsers, active development of the specification is discontinued due to performance problems with the synchronous access mechanism. As an alternative, the Indexed Database specification [16] offers a more extensive, asynchronous storage model, where data objects can be stored, indexed, and queried.

Alternatively, a set of specifications [20] defines the necessary APIs to offer Web applications a file system, where folders can be created and files can be stored. This file system lives within the browser container and is not related to the operating system's file system. By having access to a file system, Web applications can store documents for offline viewing or editing, store image galleries, etc.

All these data storage mechanisms use a security model based on the Same-Origin Policy. Each origin has its own storage container, and data-sharing between origins is not possible. Within one origin, there are no restrictions on accessing the storage containers.

The availability of application data at the client side allows Web applications to operate in an *offline* mode, as long as they do not need server-side processing. A prime example is Google's productivity suite, which includes a JavaScript-based spreadsheet application that runs entirely in the browser, only using server APIs for file storage and sharing. These kinds of applications can benefit from the *application cache*, introduced in HTML 5 [2]. The application cache allows applications to define a manifest, specifying which files the browser must cache to enable offline use. The browser transparently loads the files from the application cache when in offline mode, and updates the cached version with new files when in online mode.

3.2.2   Communication Mechanisms

The traditional script-based communication mechanism uses the XMLHttpRequest object, and was restricted to communication within the same origin. This limitation has sparked creative solutions to bypass the same-origin restriction using script tags, JSON [4] and "padding." This technique, called JSON-P, dynamically loads a new JavaScript file using a script tag, and provides the name of a callback as a parameter in the URI. The server responds with the requested data in the form of a script file, which contains an invocation of the callback, with the data in the JSON format as the argument. This not only effectively enables cross-origin communication but also introduces a severe security vulnerability, where any content can be injected into the site.

In response to this dangerous practice, the XMLHttpRequest Level 2 specification [23] makes cross-origin communication explicitly possible by implementing the Cross-Origin Resource Sharing (CORS) specification [21]. CORS allows servers to explicitly allow a wide variety of cross-origin traffic, by defining a security policy that needs to be enforced by the browser, to ensure that unsuspecting legacy servers cannot be attacked by the newly introduced client-side capabilities. Essentially, CORS ensures that the cross-origin requests that can be processed by legacy servers are restricted to the types of requests that could already be sent using a traditional HTML form. While the specification largely succeeds in this goal, some caveats still remain [7].

Next to the extension of the XHR object, several new communication mechanisms have been introduced as well. The Server-Sent Events specification [10] allows a script to keep a connection open, so it can be notified by the server in case of interesting events. Web Sockets [11] introduces a way to open a communication channel to send arbitrary data, not limited to the format of HTTP messages. Setting up a Web socket involves upgrading an HTTP channel in cooperation with the server, which is responsible for checking the origin of the party that initiates the upgrade of the channel. A third mechanism is WebRTC [1], a specification that enables real-time peer-to-peer communication between browsers. One of the primary use cases for WebRTC is audio and video chat directly in the browser, for which the HTML Media Capture specification [5] is crucial. The WebRTC protocol suite offers negotiation protocols to set up a session, even behind firewalls, using a setup server. At the time of writing, WebRTC is in full development, triggering challenges with identity management and verification [8].

3.2.3   Mobile Features

With the rise of mobile devices and their associated mobile browsers, new features are being covered in specifications as well. For example, the Geolocation API [19] offers a way to physically locate the device that is running the browser. The Vibration API [14] offers Web applications a way to trigger the vibration functionality of phones and Web notifications [22] allows a Web application to send system-level notifications to the user. In another example, the Device Orientation Events specification [3] offers Web applications a way to listen to events regarding the device orientation and movement, data that is collected from gyroscopes, compasses, and accelerometers.

The Geolocation API uses a permission model based on the host component of the origin, where the user explicitly has to grant permissions to the domain to access the location information of the device. Similarly, to receive Web notifications, the user must grant the origin permission to display system-level notifications. The other specifications mentioned here are not invasive, and do not require user permissions.

3.2.4   Registering Default Applications

Delegating content handling to other applications is a solution that has existed for a long time, and is mainly used by the *mailto*: links that invoke an email client. Whenever such a link is opened, the browser opens the system default or user-configured email client, using the parameters of the *mailto*: link.

HTML 5's *custom scheme and content handlers* [2] take this delegation to a new level, enabling Web applications to register themselves as the handler for a scheme or content type. Whenever the browser encounters a URI with a specific scheme, or a resource of a specific content type, the registered handler will be invoked. For example, a Web mail application can register itself for the *mailto:* scheme, allowing the user to directly compose a mail after clicking such a link. The possibilities are endless with whitelisted schemes such as *irc*, *sms*, *magnet*, etc. In a process similar to these schemes, a Web application can register itself as a handler for a specific content type, such as PDF documents or audio files.

3.3   Transforming the Browser into an Operating System

Modern browsers are often referred to as the *new operating system* [15, 24], a fairly inaccurate statement. However, with the rise of Google's Chromebooks, a notebook that runs on Chrome OS and Mozilla's Firefox OS, a mobile operating system that runs on top of the Gecko rendering engine, these claims may carry some truth. What we are really seeing are small operating systems that are reduced to their core tasks, such as interacting with hardware and scheduling processes, and run a single application, a browser engine. Within the browser engine, the interface layer of the operating system is loaded, which in turn supports loading additional applications and carrying out tasks.

Fig. 3.1 Firefox OS uses an
Android-based Linux kernel,
which runs the Gecko
browser engine that in turn
runs the Gaia UI, offering the
features you expect from a
mobile operating system
(Simplified depiction of the
full architecture [18])

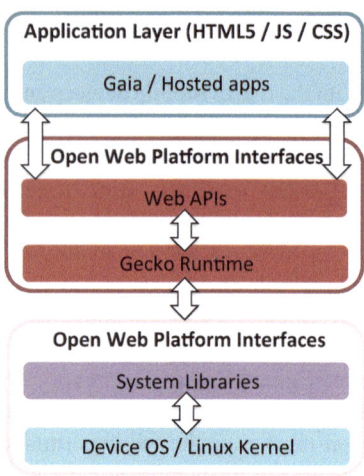

Take Google's Chromebooks for example. These custom-built notebooks have lightning fast boot times, and load Chrome OS. Chrome OS is a modified version of the Chrome browser running on top of a Linux kernel and presents the user with a familiar browser interface. Within the browser, the user can choose to browse the Web, or install any of the available applications from the Chrome Web store. These include Google's productivity suite with a text editor and spreadsheet, as well as numerous third-party applications. Chromebooks are most useful when permanently connected to the Web but also support offline use [17]. For example, applications are available offline and depending on the application, many of their features still work. For example, Google Docs automatically synchronizes documents from Google Drive allowing you to edit them offline.

An alternative to the proprietary Chrome OS is Firefox OS [18], a mobile operating system developed by Mozilla, based on a Linux core and Mozilla's Gecko browser engine. In essence, Firefox OS is an Android-based operating system, where the entire UI layer is a Web application, capable of loading other apps that are also Web applications. The architecture of Firefox OS is depicted in Fig. 3.1.

Firefox OS has three main components. The core of the operating system is called *Gonk* which consists of an Android-based Linux kernel and the hardware abstraction layer. Gonk exposes the necessary APIs to *Gecko*, the application runtime environment, which supports HTML, CSS, and JavaScript. Gecko has full access to the exposed APIs but not to the other parts of the operating system. For example, Gonk exposes the telephony system to Gecko through the *Radio Interface Layer*, controlled by the *RILd* process. The main application run by Gecko is *Gaia*, the operating systems UI layer, which implements features such as the lock screen, the home screen and the applications you expect on a smartphone. Additionally, third-party applications can be installed next to the Gaia layer, allowing users to install their own apps.

All applications that run on Firefox OS are written in HTML, CSS, and JavaScript. Since the OS runs no native applications, all system access is mediated through Web APIs, including access to the device's settings, filesystem, etc. Firefox OS strictly limits access to sensitive APIs based on trust level. Possible values are, in the order of trustworthiness. Certified apps, which are shipped with the phone, have access to the Web APIs, including sensitive APIs such as the telephony system. Privileged apps come from an authorized marketplace and have been reviewed, approved, and signed. They have access to a subset of the Web APIs but not to the sensitive APIs. The lowest trust level can only access those Web APIs that have sufficient security mitigations to be exposed to untrusted content, such as the camera or notifications API. Naturally, before access to these APIs is granted, the user has to explicitly grant the app permission upon installation time.

References

1. Bergkvist, A., Burnett, D.C., Jennings, C., Narayanan, A.: WebRTC 1.0: real-time communication between browsers. W3C Working Draft (2013)
2. Berjon, R., Faulkner, S., Leithead, T., Navara, E.D., O'Connor, E., Pfeiffer, S., Hickson, I.: HTML 5.1 specification. W3C Working Draft (2014)
3. Block, S., Popescu, A.: DeviceOrientation event specification. W3C Working Draft (2011)
4. Bray, T.: The javascript object notation (JSON) data interchange format. RFC Proposed Standard (RFC 7159) (2014)
5. Burnett, D.C., Bergkvist, A., Jennings, C., Anant, N.: Media capture and streams. W3C Working Draft (2013)
6. Dahl, D., Sleevi, R.: Web cryptography API. W3C Last Call Working Draft (2014)
7. De Ryck, P., Desmet, L., Philippaerts, P., Piessens, F.: A security analysis of next generation web standards. Tech. rep., European Network and InformationSecurity Agency (ENISA) (2011)
8. Desmet, L., Johns, M.: Real-time communications security on the web. IEEE Internet Comput. (2014)
9. Google: Google docs. http://www.google.com/docs/about/ (2014)
10. Hickson, I.: Server-sent events. W3C Candidate Recommendation (2012)
11. Hickson, I.: The WebSocket API. W3C Candidate Recommendation (2012)
12. Hickson, I.: Web workers. W3C Candidate Recommendation (2012)
13. Hickson, I.: Web storage. W3C Recommendation (2013)
14. Kostiainen, A.: Vibration API. W3C Last Call Working Draft (2014)
15. Lee, T.: The browser is the new operating system. https://www.techdirt.com/articles/20080530/0022021266.shtml (2008)
16. Mehta, N., Sicking, J., Graff, E., Popescu, A., Orlow, J., Bell, J.: Indexed database API. W3C Candidate Recommendation (2013)
17. Morris, J.: What chromebooks can do offline. http://www.zdnet.com/what-chromebooks-can-do-offline-7000027307/ (2014)
18. Mozilla Developer Network: Firefox OS architecture. https://developer.mozilla.org/en-US/Firefox_OS/Platform/Architecture (2014)
19. Popescu, A.: Geolocation API specification. W3C Recommendation (2013)
20. Ranganathan, A., Sicking, J.: File API. W3C Last Call Working Draft (2013)
21. van Kesteren, A.: Cross-origin resource sharing. W3C Recommendation (2014)
22. van Kesteren, A., Gregg, J.: Web notifications. W3C Last Call Working Draft (2013)

23. van Kesteren, A., Aubourg, J., Song, J., Steen, H.R.M.: XMLHttpRequest. W3C Working Draft (2014)
24. Wayner, P.: 10 reasons the browser is becoming the universal OS. http://www.infoworld.com/d/applications/10-reasons-the-browser-becoming-the-universal-os-230812 (2013)
25. Zalewski, M.: The Tangled Web: A Guide to Securing Modern Web Applications. San Francisco, No Starch Press (2012)

Chapter 4
How Attackers Threaten the Web

The previous chapters have shown that the Web platform and the client-side execution platform are complex environments, with several components, interaction
patterns, and policies. The variety of features offered by the Web platform enables
the development of highly sophisticated Web applications. Unfortunately, attackers
aim to abuse these features of the Web platform, attempting to perform actions in the
user's name, stealing valuable information, or just cause mayhem.

In this chapter, we will focus on the different kinds of attackers that are defined in
academic literature, and what they are capable of. Understanding these threat models
and their capabilities is crucial for fully grasping the chapters to come, as they will
flesh out concrete attacks, which require certain capabilities from an attacker.

The first part of this chapter covers the relevant academic threat models for the Web
platform. We discuss the power of the attacker in each threat model and instantiate
the threat model in a scenario applied to our example application of a social network.

Academic threat models are often highly tailored to a specific problem statement
and solution, and different reference works have slightly different definitions, which
makes comparisons of different threat models challenging. Therefore, in the second
part of this chapter, we decompose each of the presented threat models into concrete
attacker capabilities. Attacker capabilities are specific, legitimate actions that can be
performed within the Web, both by non-malicious users as well as by attackers.

4.1 Threat Models in Literature

In this section, we present the relevant academic threat models for client-side Web
security. We explain each threat model with a scenario applied to the social network
application introduced in Chap. 1.

© Philippe De Ryck, Lieven Desmet, Frank Piessens, Martin Johns 2014   33
P. De Ryck et al., *Primer on Client-Side Web Security,*
SpringerBriefs in Computer Science, DOI 10.1007/978-3-319-12226-7_4

4.1.1   Forum Poster

A *forum poster* [2] is the weakest threat model, representing a user of an existing Web application who does not register domains or host application content. A forum poster uses a Web application and potentially posts active content to the application, within the provided features. In addition, a forum poster remains standards-compliant, and cannot create HTTP(S) requests other than those he can trigger from his browser.

Our social network example allows users to post active content to the timeline. An instantiation of the *forum poster* threat model would be a malicious user posting content that contains malicious JavaScript code, also known as a cross-site scripting or XSS attack (explained in Chap. 8). When another user views the posted content, the forum poster's malicious code will be executed in the user's browser.

4.1.2   Web Attacker

The *Web attacker* [1–4] is the most common threat model encountered in papers and represents a typical attacker who is able to register domains, obtain valid certificates for these domains, host content, use other Web applications to post content to, etc. Since none of these capabilities requires a special physical location or any other typical attacker properties, every user on the Web is able to obtain them. Therefore, academic literature assumes that every other threat model except for the forum poster possesses these capabilities.

As an example of how the *Web attacker* can use his capabilities to threaten our social network, we will assume that the attacker hosts a popular image gallery with lolcats, a well-known Internet phenomenon that originated on postcards in the 1800s. However, next to the image gallery code, the attacker's page includes HTML code that will send a background request to the social network, resulting in the provided content being posted to the user's message board. Whenever the victim visits the image gallery, while still being signed in to the social network, e.g., in another browser tab, the background request will result in the attacker-provided content being posted to the victim's timeline. This attack is known as a cross-site request forgery or CSRF attack (explained in Chap. 6).

4.1.3   Gadget Attacker

A *gadget attacker* [1, 3] is a more powerful version of the Web attacker, where the attacker hosts a component that is wilfully integrated into the target application. Popular examples are JavaScript libraries, such as JQuery; analytics code, such as Google analytics; or widgets, such as Google Maps. The gadget attacker is extremely relevant in the context of code isolation for mashups or complex, composed sites, which integrate content from multiple stakeholders with varying trust levels.

Our social network supports the integration of third-party applications, such as weather gadgets, stock tickers, games, etc. The presence of a gadget attacker is inherent to this design. For example, a malicious developer can create a legitimate-looking weather gadget, allowing the users to add weather information to their main page. Since the gadget is JavaScript code running in the user's browser, it can perform all kinds of operations in the background, such as reading and leaking the user's private messages, changing the appearance of the Web site, also known as defacing, or actually sending requests in the user's name.

4.1.4   Related-Domain Attacker

A *related-domain attacker* [4] is an extension of the Web attacker, where the attacker is able to host content in a related domain of the target application. A common case of a related-domain attacker is when the attacker is able to host content on a sibling or child domain of the target application, e.g., for the Web sites of different departments within a company.

Our social network sells personal spaces to commercial business owners, allowing them to represent their business within the social network. Such personal spaces are hosted as a subdomain of the social network, e.g., *mybusiness.oursocialnetwork.com*. A related-domain attacker purchasing such a personal space can abuse specific features reserved for child domains. For example, the attacker can read cookies that are set for the *oursocialnetwork.com* domain and can create his own cookies that will be sent to the social network's main page. The latter can be used to mount a session fixation attack, as explained in Chap. 7.

4.1.5   Related-Path Attacker

A *related-path attacker* is another extension of the Web attacker, and represents an attacker that hosts an application on a different path than the target application, but within the same origin. This scenario occurs, e.g., within the Web hosting of Internet Service Providers (ISPs), which often offer each of their clients a Web space under a specific path, all within the same origin. Academic papers aptly describe this attacker [6] and its conflicts with the Web's security model, albeit without giving it an explicit name.

For example, if we would host the commercial spaces in our social networking example under a single subdomain, but with different paths, we would allow these individual spaces to circumvent the Same-Origin Policy. Since they would all reside within the same origin, they would share the same cookies, have access to each other's origin-constrained resources, and would be allowed to reach into each other's frames.

4.1.6   Passive Network Attacker

A *passive network attacker* [7] is considered to be an attacker who is able to passively eavesdrop on network traffic but cannot manipulate or spoof traffic. A passive network attacker is expected to learn all unencrypted information. In addition, a passive network attacker can also act as a Web attacker, for which no specific requirements are needed.

One common example of a passive network attack is an attacker eavesdropping on unprotected wireless communications, which are ubiquitous, thanks to publicly accessible Wi-fi networks and freely available hotspots. If the passive network attacker observes any unencrypted network traffic between the victim and our social network, he is not only able to extract personal information, but also important metadata, such as the user's cookies. Such cookies typically contain a session identifier, which can in turn be used to mount a session hijacking attack, as explained in Chap. 7.

In 2013, whistleblower Edward Snowden [8] revealed that intelligence services across the globe have powerful traffic monitoring capabilities. These *pervasive monitoring* capabilities are passive network attacks, albeit on a very large scale compared to the traditional passive network attacker. In response to the Snowden revelations, the IETF has drawn up a best practice, stating that specifications should account for pervasive monitoring as an attack [5].

4.1.7   Active Network Attacker

An *active network attacker* [1, 2, 7] is considered to launch active attacks on a network, for example by controlling a DNS server, spoofing network frames, offering a rogue access point, etc. An active network attacker has the ability to read, control and block the contents of all unencrypted network traffic. An active network attacker is generally not considered to be capable of presenting valid certificates for HTTPS sites that are not under his control, unless by means of attacks such as SSL stripping (covered in Chap. 5).

In our example, the active network attacker configures his own wireless access point and chooses the name of popular hotspots provided by the local ISP. Unsuspecting users are used to connecting to these networks, and visit our social network using the attacker's wireless network. The attacker can now not only read unencrypted information but also change the content posted by the user and modify the responses sent to the user. In addition, should our social network use encrypted connections, the attacker can attempt to break or downgrade the level of security, potentially allowing him to read or manipulate even the encrypted traffic.

4.2   Threat Models as Concrete Attacker Capabilities

The threat models presented in academic literature are highly tailored to a specific problem statement and solution, and different reference works have slightly different definitions, making general reasoning with threat models, or even comparisons of threat models within the Web platform difficult.

Therefore, we break down the academic threat models into concrete attacker capabilities, which entail an action that can be performed by an attacker on the Web, as shown in Table 4.1. Generally, these actions are legitimate operations within a certain context, and are often performed by users, developers, and companies in various legitimate scenarios. However, by using a capability to exploit a certain feature of the Web, an attack can be mounted against a victim.

One example is a *Web attacker* that *registers an available domain* with a *valid certificate for the domain*. By using his capabilities to perform these legitimate actions within the Web platform, the attacker has now constructed a phishing setup, where he will trick unsuspecting users into entering their credentials into a fraudulent authentication form.

Essentially, attacker capabilities are fundamental operations that can be carried out within the Web. Naturally, the exact set of capabilities an attacker possesses depends on his position in the Web ecosystem. An attacker that merely hosts a Web site under his own domain will not have the capability to eavesdrop on a user's local network traffic, but an attacker sitting next to the user, using the same wireless network might.

In the remainder of this section, we will cover each of the identified attacker capabilities, carefully expressing the ways of achieving the capability, and the associated power.

4.2.1   Send Requests to an Application

A user on the Web sends requests to applications and receives responses in return. This is the basic behavior of the Web and is also available to attackers. Depending on the application, the authentication and authorization infrastructure, an attacker might have access to public resources only, or resources deep within the application. Note that when requests are sent by a browser, they should be standards-compliant and cannot deviate from the implemented protocols, but that a user who controls his client machine can issue nonstandards-compliant requests as well.

4.2.2   Register Own Domain

Any Internet user, including an attacker, is able to register a currently unregistered domain. The procedure commonly requires the payment of an annual service fee to

Table 4.1 An overview of academic threat models, decomposed into fine-grained attacker capabilities. The grouping of the capabilities is based on their specific characteristics

	Forum poster	Web attacker	Gadget attacker	Related-domain attacker	Related-path attacker	Passive network attacker	Active network attacker
Send requests to an application	★	★	★	★	★	★	★
Register own domain		★	★	★	★	★	★
Host content under own domain		★	★	★	★	★	★
Respond to requests from own domain		★	★	★	★	★	★
Register valid cert for own domain		★	★	★	★	★	★
Manipulate target's domain-based data			★	★	★	★	★
Manipulate target's client-side context			★		★	★	★
Eavesdrop on network traffic						★	★
Generate network traffic							★
Intercept and manipulate network traffic							★

a registrar, who is licensed to hand out subdomains for the specific parent domain (e.g., registering *example.org.uk* would be authorized by the licensee for the *org.uk* domain). Getting control of an already-registered domain is not possible, unless by court order or by snatching it away after its previous owner had let it expire.

4.2.3   Host Content Under Own Domain

Hosting content under an own domain is a basic capability for Web developers, and thus also available to any Web user with malicious intentions. Having a victim visit the attacker-controlled content is easily achieved, e.g., by posting links to social networking sites, offering interesting content, etc. Once a victim visits the attacker content, the attacker-provided code runs within the victim's browser, constrained within its origin.

Setting up hosted content is fairly straightforward, especially using hosting services from a provider, in exchange for a small service fee. Well-organized attackers issue such payments using stolen credit cards or entire identities, making it difficult to track and prevent such transactions.

4.2.4   Respond to Requests from Own Domain

Whenever an attacker controls a Web application, he can send arbitrary responses to client requests. Note that this is an explicit capability, because an attacker is not necessarily bound by HTTP specification or underlying Web serving software, and can attempt to exploit vulnerabilities at the client side.

4.2.5   Register a Valid TLS Certificate for Own Domain

Setting up secure connections using HTTPS requires a valid certificate, otherwise the browser will generate disconcerting warnings to the user. An attacker, or anyone for that matter, can apply for a certificate for a given domain name, as long as the identity verification checks succeed. Passing simple ownership validation is straightforward, and is, for example, used when attackers register a domain name that resembles the target domain, thereby obtaining a valid certificate for the fraudulent domain, hoping to impersonate the target domain to victims. Obtaining certificates for non-attacker controlled domains should be impossible, but depending on the verification process and/or gullibility of the administrators, an attacker might succeed in obtaining a certificate for a domain that is not controlled by the attacker. Once an attacker obtains a valid certificate for a domain, he is able to impersonate a legitimate server for this domain over a secure connection, allowing him to intercept and manipulate network traffic.

4.2.6   Manipulate Target's Domain-based Data

Depending on the situation, an attacker may be capable of manipulating the target's domain-based client-side data. This includes any data that is associated with the parent domain, and becomes accessible through domain relaxation using the *document.domain* property. The most common example are cookies assigned to the parent domain, which are valid for all subdomains. Another example are client-side storage facilities using origin-based constraints that may be accessible to multiple applications.

4.2.7   Manipulate Target's Client-Side Context

An attacker that is capable of manipulating the target application's client-side context, essentially bypasses any constraints enforced by the Same-Origin Policy. The client-side context is extremely important, as it gives access to the DOM-tree and to any client-side resource constrained by the origin. An attacker that has this capability can run code within the target's application origin, either directly on the page or on a related page within the same origin.

4.2.8   Eavesdrop on Network Traffic

In certain circumstances, an attacker might be able to eavesdrop on network traffic from legitimate users. One common case is an attacker using the same wireless network, who is able to receive all transmitted data. Alternatively, attackers on a wired network may also be able to see a certain fraction of network traffic. By eavesdropping on the network, an attacker can gather valuable information, which can be used to escalate an attack against a user or application. Note that eavesdropping can be done in a completely passive way, preventing detection by the network infrastructure or monitoring software.

4.2.9   Generate Network Traffic

In addition to eavesdropping on traffic, a network attacker can also generate new traffic with spoofed parameters. For example, if an attacker sees a browser making a request for a certain resource, he can generate a response that seems to come from the target server, and send it before the actual response reaches the browser. This allows an attacker to provide malicious content to legitimate client applications, potentially compromising the application or even the client machine.

One example of how a traffic generation attack can be conducted was uncovered by the Snowden [8] revelations. The NSA's QUANTUMINSERT program monitors the network close to the target and detects a request going out to a specific application. Upon detection of the request, a response with a fake page is injected into the network, and will very likely reach the victim's browser before the original response. The fake page can be used to redirect the victim to a malware server, part of the NSA's FOXACID program.

4.2.10   Intercept and Manipulate Network Traffic

A third network-based capability is to intercept and manipulate network traffic, conducting a full Man-in-the-Middle (MitM) attack. An attacker sitting in the path between a user and a server can intercept and inspect every request and response. In addition, he can modify them to his wishes, potentially compromising the client-side or server-side context of the Web application. Carrying out a MitM attack is far from stealthy and can be detected by monitoring software. In addition, whenever HTTPS is used, an attacker has to obtain an appropriate certificate for the client and/or server, in order to prevent alarm bells from going off.

4.3   Conclusion

Table 4.1 shows how the academic threat models can be expressed in concrete attacker capabilities. The table clearly shows the differences in power between the threat models. A *forum poster* can merely send requests to an application, while an *active network attacker* has much more power. Note that while an *active network attacker* may be significantly stronger, he is also capable of performing the same actions as the *forum poster*. By using concrete capabilities, the requirements for an attack can be expressed more finely. For example, in a session hijacking attack (Chap. 7), an attacker will eavesdrop on an insecure network to steal the session identifier, which is in turn used in a legitimate request to the application. This could be executed by a *passive network attacker*, or an attacker with the capability to *eavesdrop on network traffic* and *send a request to the application*.

The academic threat models covered in this chapter lie at the basis of every attack covered in the subsequent chapters. Attacks often have multiple attack vectors, which require a different set of concrete capabilities. Choosing a concrete attack vector depends on the power of the attacker, and the exploitability of the application vulnerabilities. The nuances between concrete attacker capabilities will help in understanding these attacks, their attack vectors, and their countermeasures.

References

1. Akhawe, D., Barth, A., Lam, P.E., Mitchell, J.C., Song, D.: Towards a formal foundation of web security. In: Proceedings of the 23rd IEEE Computer Security Foundations Symposium (CSF), pp. 290–304 (2010)
2. Barth, A., Jackson, C., Mitchell, J.C.: Robust defenses for cross-site request forgery. In: Proceedings of the 15th ACM Conference on Computer and Communications Security (CCS), pp. 75–88 (2008)
3. Barth, A., Jackson, C., Mitchell, J.C.: Securing frame communication in browsers. Commun. ACM **52**(6), 83–91 (2009)
4. Bortz, A., Barth, A., Czeskis, A.: Origin cookies: Session integrity for Web applications. Web 2.0 security and privacy (W2SP) (2011)
5. Farrel, S., Tschofenig, H.: Pervasive monitoring is an Attack. RFC Best Current Practice (RFC 7258) (2014)
6. Jackson, C., Barth, A.: Beware of finer-grained origins. Web 2.0 Security and Privacy (W2SP) (2008)
7. Jackson, C., Barth, A.: Force HTTPS: Protecting high-security web sites from network attacks. In: Proceedings of the 17th International Conference on World Wide Web (WWW), pp. 525–534 (2008)
8. The Guardian: Edward Snowden. http://www.theguardian.com/world/edward-snowden (2013)

Chapter 5
Attacks on the Network

The previous chapters of this book have introduced the Web platform and the different kinds of attackers that are present within the Web. This chapter covers a first set of attacks, executed at the network level, somewhere between the user's browser and the Web application's server.

Attacks on the network level may be further away from the user, but that does not take away their power. Successfully executing a network attack not only allows an attacker to eavesdrop or manipulate network traffic, but also to use these capabilities as a stepping stone towards many other attacks, as you will learn in the later chapters.

In this chapter, we cover three different kinds of network attacks. First, we discuss eavesdropping attacks, where an attacker listens in on the network traffic being sent over the wire or through the air. Next, we focus on man-in-the-middle (MitM) attacks, where the attacker intercepts and manipulates the traffic. Finally, we focus on attacks on Hypertext Transfer Protocol Secure (HTTPS), which already uses Transport Layer Security (TLS) to add confidentiality, integrity, and entity authentication.

5.1 Eavesdropping Attacks

In an eavesdropping attack, an attacker listens in on other users' network traffic, such as Domain Name System (DNS) queries, HTTP requests and responses, etc. By eavesdropping on their network traffic, an attacker is not only able to learn sensitive and personal information such as credit card info, financial means, usernames, passwords, contents of email messages, etc., but can also listen in on important Web metadata, such as session identifiers or supposedly secret cookies. Obtaining any of this information is not only directly harmful to the user but also enables the attacker to escalate the attack, for example, through *session hijacking* (covered in Chap. 7).

Eavesdropping attacks are extremely relevant in the modern Web, especially because of the numerous wireless networks, to which users connect with their mobile devices or laptops. Many of these networks are unprotected or easily spoofed by an attacker. Additionally, with the revelations of Snowden [50], it has become clear that state-sponsored eavesdropping occurs on a large scale, scooping up every piece of unencrypted information that is encountered.

© Philippe De Ryck, Lieven Desmet, Frank Piessens, Martin Johns 2014   43
P. De Ryck et al., *Primer on Client-Side Web Security,*
SpringerBriefs in Computer Science, DOI 10.1007/978-3-319-12226-7_5

5.1.1 Description

The goal of an eavesdropping attack is to obtain traffic that is sent over the network. The way of executing an eavesdropping attack depends on the network under attack. For example, eavesdropping on airborne signals, such as WiFi, radio, or cellular, only requires an antenna in the proximity of the network. Eavesdropping on a switched wired network requires some interference, for example by running an Address Resolution Protocol (ARP) spoofing attack. Eavesdropping can also occur at intermediaries within the network infrastructure, for example at an Internet service provider (ISP), a proxy server, or a Tor [12] exit node. Even higher up in the network, an attacker can eavesdrop on the backbone traffic, with submarine taps on fiber optic cables [4] as an extreme example.

Technically, running an eavesdropping attack is fairly straightforward. As an illustration, the browser add-on *Firesheep* [8] enables a user to eavesdrop on a WiFi network, abusing obtained session identifiers to perform a session hijacking attack with one point-and-click operation. Alternatively, software tools such as Subterfuge [52] and dedicated devices such as the *Pineapple* [23] make collecting sensitive information a straightforward task. Eavesdropping on a wired, switched network is also possible with a wide variety of freely available tools, such as *Ettercap* [15] or *dsniff* [49].

Essentially, eavesdropping attacks will always be possible, especially with the evolution towards wireless networks. However, the real problem with eavesdropping is the huge amount of information that is transmitted in the clear, making the physical access to the network signals the only barrier to overcome.

5.1.2 Mitigation Techniques

The main approach to protect network traffic against eavesdropping attacks is to deploy security protocols, effectively protecting the data being sent over the network. The use of network-specific data link-layer security protocols, such as WiFi protected access (WPA) [54] and Extensible Authentication Protocol (EAP) [1] can effectively help in mitigating a local eavesdropping attacker but does not protect the traffic against eavesdropping beyond the local network.

An approach offering end-to-end security is using HTTP deployed over TLS [11], where TLS is aimed at offering confidentiality, integrity, and entity authentication. The confidentiality property effectively eliminates the usefulness of the data in transit to an eavesdropping attacker. As the next section will explain, the integrity and entity authentication properties mitigate active network attacks. Note that TLS uses certified public keys to establish entity authentication, but that confidentiality against passive network attackers can already be achieved using self-certified keys. Such a configuration is, however, not recommended for general use due to the weaker security guarantees.

The TLS protocol itself is undergoing constant revision and its security is the topic of ongoing cryptographic research. Recent examples are the discovery of timing attacks against the CBC mode of operation [2], allowing the extraction of cookies from the encrypted stream, and the identification of weaknesses in the RC4 algorithm [3], supporting the common belief that RC4 should be considered broken. Reacting to these results and other similar research output, the TLS working group within the Internet Engineering Task Force (IETF) is currently considering new cipher suites, for example based on the *ChaCha20* cipher [29].

In addition to new ciphers, TLS deployment is also an important factor to consider. Older versions of TLS remain in use for a considerable period, even after the introduction of newer versions with better security features. For example, the attacks mentioned in the previous paragraph have already been countered via authenticated encryption cipher suites that were defined as a part of TLS 1.2 [10], but so far version 1.2 has not seen very widespread deployment. However, in response to these attacks, deployment roadmaps for TLS 1.2 have been accelerated by browser vendors and the open-source security community.

Another standardization proposal focuses on a new best current practice (BCP) for the use of TLS on the Web [48], aiming to aid Web application developers and administrators in the use of TLS. One example strategy that is advocated is the achievement of *perfect forward secrecy* (PFS), which guarantees that previously recorded TLS sessions cannot be deciphered by learning the server's private key at a later point in time. While PFS has previously been little deployed due to an impact on performance and requiring less commonly used cipher suites, it has come into the spotlight again due to private keys leaking out, for example if someone hacks a Web server or if server units are decommissioned inappropriately.

Finally, a large effort is being spent on the development of the successor to HTTP, HTTP/2.0 [7] based on Google's SPDY protocol [6]. While initially SPDY was proposed to always run over TLS, that position was somewhat watered down in HTTP/2.0, mainly due to interference with middleboxes, such as Web caches or HTTP proxies that need to be able to see some HTTP metadata (e.g., headers) to function. Nonetheless, HTTP/2.0 will be far more likely to be deployed running over TLS, since the *application layer protocol negotiation* TLS extension [19] offers the most efficient upgrade path, with the fewest additional network round trips. Additionally, discussions to allow clients and servers to make use of TLS even for HTTP (i.e., non-HTTPS) URIs when using HTTP/2.0 are in process [37].

5.1.3   State of Practice

While TLS effectively tackles network attacks, it is not yet widely deployed, although adoption is growing. As an indication of the current deployment state of TLS, a monitoring site [53] reports that approximately 34 % of the top 10-million Web sites are using TLS with certificates issued by a recognized certificate authority (CA). Recently, Google has announced an *HTTPS Everywhere* initiative, encouraging the deployment of TLS. As a part of this initiative, Google is starting to use HTTPS as a signal in their ranking algorithm [5].

Next to the limited adoption, older and less secure versions of TLS that are deployed, are rarely upgraded to the latest version, leaving a trail of inadequate legacy implementations across the Web. The SSL Pulse project [40] reports that of the 152,733 surveyed TLS sites in July 2014, only 28.3 % had a secure TLS deployment.

While the specific reasons for the slow adoption of TLS are hard to pinpoint, potential candidates are its (antiquated [28]) reputation imputing significant performance impact, the difficulty of managing and deploying certificates, potential interference or incompatibility of the encrypted traffic with middleboxes, such as proxies and caches, and general ignorance of Web application operators. Additionally, TLS is often used incorrectly, which may be attributed to relatively hard-to-use application programming interface (APIs) and incorrectly configured trust-roots.

5.1.4 Best Practices

The best practice to protect network traffic against eavesdropping attacks is to deploy TLS everywhere (see the best practices of Sect. 5.2 for additional information), achieving confidentiality for all data and metadata sent between the user's browser and the Web application. Additionally, by selecting cipher suites offering perfect forward secrecy, the encrypted data is even protected against an attacker who learns the private key in the future. The IETF's BCP document exclusively covers the best practices with regard to TLS deployment [48], as does the work of Ivan Ristić [42, 43]. You should also test your site's certificate and configuration using tools such as Qualys SSL Labs [39].

5.2 Man-in-the-Middle Attacks (MitM)

An MitM is an active network attack, where the attacker positions himself in the network, between the victim and the targeted Web application. This position not only allows the attacker to inspect all traffic that is sent between the victim and the target application but also allows modification of the traffic. Such a compromise gives the attacker full control over the user's actions, with potentially disastrous effects. Note that there are also 'legitimate' use cases for performing an MitM attack, such as ISPs injecting advertisements into HTTP responses, or corporations deploying a Web content filter responsible for filtering unwanted or harmful content.

MitM attacks are more sophisticated than eavesdropping attacks and occur frequently on the Web. Little is known about MitM attacks being carried out by small-scale attackers, but they do occur on larger scales, such as for state-sponsored censorship as seen in the Middle East, China, etc. Similarly, the same technology is used for scenarios where user consent is given, such as companies that deploy content filtering on their own networks, as a perimeter security measure [26].

5.2.1 Description

The goal of an MitM attack is to be able to inspect and manipulate the victim's network traffic. This allows an attacker to modify legitimate transactions, carry out actions in the user's name, compromise files that are being sent to and from the victim and many more. TLS-secured connections with validated certificates are designed to withstand MitM attacks, but flaws in the supporting systems may allow for subtle attacks to be carried out anyway. These flaws are caused by misplacement of trust in certain parties or by placing the decision-making burden on the user.

Actually, becoming an MitM in the network can be achieved at many levels. An attacker can physically place a machine in the network path, forcing the data to flow through this machine, or can manipulate the network's parameters, to act as a gateway on the logical level, for example through ARP poisoning attacks. The technical details on becoming an MitM are less important, but the impact of an MitM attack on the Web is. Once an attacker positions himself in the middle, inspecting and manipulating traffic becomes straightforward.

Traditionally, TLS is deployed to prevent eavesdropping and MitM network attacks, since it offers confidentiality, integrity, and entity authentication. The confidentiality and integrity effectively prevent an attacker from modifying any network traffic, while the entity authentication property ensures that the involved parties are who they claim they are, thereby preventing an MitM attack within the TLS connection. Even though TLS is designed to counter MitM attacks, in reality, they remain possible for several reasons.

In 2009, Moxie Marlinspike argued [32] that users visiting a secure Web application probably will not type the *https://* part of the URI manually, meaning that the initial request will be made over HTTP. Typically, Web applications then redirect the user towards the correct HTTPS URI, causing a transition from HTTP to HTTPS. Exactly this transition can be exploited by an attacker that sits between the victim and the target application, causing the downgrade of the connection from HTTPS to HTTP, which is called an *SSL Stripping* attack [33].

Second, the entity authentication in TLS is based on private/public key pairs, of which the public key is verified by a CA, which is part of the public key infrastructure (PKI). Unfortunately, any CA in the Web's PKI can issue a certificate for any Web site, since no tightly bound name constraints are offered, in spite of the availability of the technology [9]. With approximately 57 trusted root CAs in a modern browser, any Web site is vulnerable to an attack with fraudulent but verified certificates being issued.

Third, whenever an invalid certificate is encountered by a browser, the burden of the security decision is placed on the user. Regardless of whether the invalid certificate is caused by an expired expiration date, or a complete mismatch with the targeted Web site, browsers show scary warnings, asking the user to decide whether to trust the site or not. Since users also encounter these warnings for legitimate sites, a simplistic MitM using an invalid certificate has some chances of success.

A fourth degradation of the CA system in TLS comes from the deliberate MitM devices, deployed by enterprises and large organizations with the goal of filtering inbound and outbound Web traffic. Reasons to deploy such filtering mechanisms go from offering protection, for example with a Web application firewall (WAF), to preventing employees from accessing sites that are deemed inappropriate, such as social networking applications. The problem with such devices is that in order to perform an MitM attack over secure connections, they have to either install their own certificate on a user's machine, or they have to obtain a valid certificate for every TLS-protected Web site on the Web. The former is a configuration hassle, which only works if you control all the client-side devices as well, and the latter seems impossible. Unfortunately, the system does not prevent collaboration between CAs and vendors of MitM devices [51], thereby harming the trust placed in the system.

Finally, the trust placed in CAs is easily abused when a CA is compromised. For example, the hacking of DigiNotar [38] resulted in the issuing of fraudulent certificates, allowing MitM attacks on secure connections to Yahoo, Mozilla, WordPress, and the Tor project. The trusted roles of CAs can even be further compromised by government coercion to issue fraudulent certificates. This strategy is believed to be a common practice in non-democratic countries [47], but recent revelations show that this practice is widely deployed by secret agencies across the world [34].

The essence of the problem with MitM attacks, especially against TLS connections, is the misplaced trust in the system on the one hand, and the burden of the security decision on the user on the other. Clearly, blindly trusting every root CA in the world has been proven to be a bad idea, and typical Web users are not capable of making technical decisions about trusting a certificate or not.

5.2.2 Mitigation Techniques

For long, the main mitigation for SSL stripping attacks has put the burden on the user, who should detect the presence of the lock icon to indicate a secure connection. One technological solution is provided by HTTPS Everywhere [14] browser add-on, which forces the use of HTTPS on sites that support it. By forcing the use of HTTPS, SSL stripping attacks are effectively mitigated, since a direct HTTPS connection will be made. The research proposal *HProxy* [36] prevents SSL stripping attacks by leveraging the browser's history to compose a security profile for each site, and validating any future connection to the stored security profiles. This approach effectively detects and prevents SSL stripping attacks without server-side support and without relying on third-party services. Finally, the Force HTTPS research proposal [27] has resulted in *HTTP Strict Transport Security (HSTS)* [24], which allows a server to require that browsers supporting HSTS can only connect over HTTPS, effectively thwarting any SSL stripping attack. A server can enable HSTS protection by including a `Strict-Transport-Security` response header, declaring the desired lifetime for the HSTS protection. One caveat to HSTS being implemented as a response header, is the first contact with a site, when it is unknown whether an HSTS policy applies or not. This issue has been addressed by modern browsers, including a predefined list of HSTS-enabled sites, effectively avoiding an initial HTTP connection.

Mitigation techniques against MitM attacks on TLS focus on determining the trustworthiness of the presented certificate. Certificate transparency (CT) [30] aims to maintain a public, write-only log of issued certificates so that either user agents or auditors can detect fraudulent certificates. This would require a user agent to query the log during the TLS handshake, and auditors can query the log offline, to check for certificates being unexpectedly issued for one of their sites.

A second approach is based on detecting discrepancies between the currently presented certificate, and previously seen certificates, a technique called *certificate pinning* or *public key pinning*. While this approach requires the first connection to be secure, it effectively enables the detection of unexpected future updates. This approach is implemented in the Certificate Patrol browser add-on [35], and proposals to achieve this at HTTP, TLS, or other layers have been made [16, 24]. Note that public key pinning does not require a CA-signed certificate and is compatible with self-signed certificates. Alternatively, Google has taken the approach of hardcoding the certificate fingerprints of Google-related TLS certificates, allowing Google Chrome to detect a potential MitM attack, even with a fraudulent certificate issued by a CA. Naturally, controlling both the services and the client platform is a key to the success of this approach.

Several proposals for alternate schemes to verify certificates have been made and evaluated [22], but the standardization work on CT seems to be most likely to gain widespread support, which is required for it to become an effective mitigation technique. In addition to CT-like approaches, DNS-based Authentication of Named Entities (DANE) [25] leverages the security of Domain Name System Security Extensions (DNSSEC), thereby avoiding the name constraint problem that enables MitM attacks. DANE is perhaps less suited for the Web due to the current lack of deployment of DNSSEC and a corresponding lack of a well-defined transition path from today's PKI to a DANE-based PKI.

5.2.3   State of Practice

Currently, a large part of the Web still transfers content over HTTP, making an MitM attack trivial. In the past few years, major sites have started to switch TLS on by default, which has even increased after the revelations about pervasive monitoring. Adoption of HSTS is still in its early stages, but our July 2014 survey of the Alexa top 10,000 domains shows that 388 have already sent an HSTS header. Similarly, Chromium's predefined list of HSTS-enabled sites counts 438 entries.

A recent study [26] has discovered that forged certificates do occur in the wild. Of the 3,447,719 real-world TLS connections, at least 6845 (0.2 %) used a forged certificate. The authors attribute these forged certificates to adware, malware, and security tools such as antivirus software, parental controls, and firewalls.

5.2.4   Best Practices

Best practices for avoiding MitM attacks on TLS-secured connections are hard to give, since preventing such attacks is entirely the goal of TLS. A good resource on secure TLS deployments is the OpenSSL Cookbook [42], which discusses current best practices, such as choosing a sufficiently strong key, fully encrypting your entire site, enabling forward secrecy, etc. Additionally, if you deploy TLS, you should configure HSTS as well.

In the meantime, experts are still debating on the future path for more secure deployments, where the publication of DANE TLSA records through DNSSEC, choosing a CA that supports CT and enabling public key pinning, are likely to be good practices.

Browsers can provide better TLS error handling, clearly indicating which errors are severe, and which are benign. Fortunately, improving the TLS experience on the user side seems to be an ongoing effort for browser vendors.

5.3   Protocol-level Attacks on HTTPS

Once a secure connection has been initiated, without an MitM being present, the transmitted content should be secure. However, sophisticated attacks on HTTPS and TLS protocols have been able to extract data from a secure connection, or to inject data into the stream. Luckily, these attacks are largely mitigated in upgraded versions of the TLS protocol. Nonetheless, actively scrutinizing and repairing security protocols remains essential for network-level security.

In this section, we provide an overview of the most relevant attacks on TLS, followed by an overview of the state of practice. There is no detailed discussion of mitigation techniques, as there are no dedicated techniques to mitigate these protocol attacks, besides keeping up with the latest version. Whenever a flaw in the protocol or tool is detected, it is either already patched in the latest version, or a solution will be released almost immediately. Take for example the Heartbleed vulnerability (covered below), for which a patch was available when the vulnerability was publicly disclosed [45].

	The Heartbleed Vulnerability
1	
2	In 2014, the *Heartbleed* vulnerability [46] was discovered, gaining plenty
3	of media attention and causing widespread panic. The vulnerability allowed
4	an attacker to extract arbitrary memory information from the server which
5	could include usernames, passwords, and the private keys belonging to the
6	TLS certificate. The Heartbleed vulnerability has severe consequences, as it
7	impacts the entire TLS deployment of a server, and not a single TLS session
8	of a single user.

9   The Heartbleed vulnerability was caused by an implementation bug in
10 OpenSSL, and has been quickly mitigated by a security patch. While this
11 vulnerability is not related to client-side Web security, we covered it here for
12 completeness. The important lesson from the Heartbleed vulnerability is that,
13 even when everything is designed to be secure, vulnerabilities will always
14 remain. In the case of Heartbleed, a large part of the vulnerable servers has
15 been patched, but as usual, a fraction remains vulnerable to this attack, aptly
16 illustrated by the theft of 4.5 million patient records from a US hospital [31].
17 Pushing updates across all servers and applications on the Web is one of the
18 core challenges in securing the Web.

Many of the attacks discussed below are effectively mitigated in TLS 1.2 [10], and as a result, the adoption of the new version is accelerating. Additionally, recent standardization activities focus on the development of TLS 1.3 within the IETF's TLS working group [11]. TLS 1.3 aims to counter all known TLS attacks, and will use more modern cipher suites, deprecating the vulnerable ones. Similarly, the development of HTTP/2.0 [7] will counter attack abusing header compression, such as the CRIME attack [44].

In addition to work on TLS, HTTP authentication schemes are receiving renewed attention, with proposals aiming to overcome the problems with HTTP Basic and Digest authentication [18], such as lack of control over the user interface, lack of a logout function, and the clear text or hashed transmission of the user credentials. One proposal is *HTTP Origin Bound Authentication* (*HOBA*) [17], which aims to provide a digital signature challenge-response mechanism to perform HTTP-based authentication. For completeness, we describe several attacks below.

5.3.1 Overview of Attacks

Client Authentication

A lesser-known feature of HTTP and TLS protocols is client authentication, where HTTP offers the Authorization header (See Chap. 2), and TLS offers the ability to use client certificates. Important use cases of TLS client authentication are WebDAV [21] deployments and server-to-server communication. A vulnerability in the TLS renegotiation procedure allowed the injection of plaintext into the channel, but confidentiality was never threatened. This attack has been quickly mitigated [41] and has seen reasonably good deployment since its release.

Lucky-13

The *Lucky-13* attack [2] is a side-channel attack against the *MAC-then-encrypt* scheme used in TLS for CBC cipher suites. Since Lucky-13 is a network timing attack, it requires the attacker to be nearby in the network and to issue tens of thousands of requests to the target application, in order to extract a plain text value, such as a session cookie.

BEAST

The *Browser Exploit Against SSL/TLS (BEAST)* [13] attack uses active scripting in the browser, in conjunction with a colluding intermediary, to exploit a CBC vulnerability, allowing the decryption of network traffic. BEAST is the first practical implementation of a previously known vulnerability and has already been mitigated in TLS 1.1.

CRIME and BREACH

The CRIME [44] and the recent BREACH [20] attack exploit compression at HTTP or TLS layers, allowing an attacker to guess plain texts based on compression ratios. These attacks can be prevented by turning off compression within TLS, the default configuration; although, turning off compression in HTTP might be less practical.

RC4 Attacks

Recent attacks against the use of RC4 [3] target sensitive plain text in fixed positions within the cipher text. Since cookie headers are often easily located, this attack impacts real-life deployment scenarios.

5.3.2   *State of Practice*

The SSL Pulse [40] project analyses the use of TLS on the most popular Web sites. Its July 2014 survey of 152,733 sites shows the following statistics:

- 7,000 sites (4.6 %) still support insecure renegotiation
- 114,497 sites (75.0 %) are vulnerable to the BEAST attack
- 14,542 sites (9.5 %) support TLS compression and are vulnerable to the CRIME attack
- 45,260 sites (29.6 %) still use RC4 when modern browsers connect
- 777 sites (0.5 %) remain vulnerable to Heartbleed

These numbers illustrate the long lifetime of legacy systems on the Web as a pressing problem. By not upgrading TLS deployments to the latest versions of both protocol and software, these sites remain vulnerable to well-known attacks, which are actively exploited in the wild.

References

1. Aboba, B., Simon, D., Eronen, P.: Extensible authentication protocol (EAP) key management framework. RFC Proposed Standard (RFC 5247) (2008)
2. AlFardan, N.J., Paterson, K.G.: Lucky thirteen: breaking the TLS and DTLS record protocols. In: Proceedings of the 34th IEEE Symposium on Security and Privacy (SP) (2013)
3. AlFardan, N., Bernstein, D.J., Paterson, K.G., Poettering, B., Schuldt, J.: On the security of RC4 in TLS and WPA. In: Proceedings of the 34th IEEE Symposium on Security and Privacy (SP) (2013)
4. Associated Press: New nuclear sub is said to have special eavesdropping ability. http://www.nytimes.com/2005/02/20/politics/20submarine.html?_r=0 (2005)
5. Bahajji, Z.A., Illyes, G.: Https as a ranking signal. http://googlewebmastercentral.blogspot.be/2014/08/https-as-ranking-signal.html (2014)
6. Belshe, M., Peon, R.: SPDY protocol. IETF Internet Draft (2012)
7. Belshe, M., Thomson, M., Melnikov, A., Peon, R.: Hypertext transfer protocol version 2.0. IETF Internet Draft (2014)
8. Butler, E.: Firesheep. http://codebutler.com/firesheep (2010)
9. Cooper, D., Santesson, S., Farrell, S., Boeyen, S., Housley, R., Polk, W.: Internet X.509 public key infrastructure certificate and certificate revocation list (CRL) profile. RFC Proposed Standard (RFC 5280) (2008)
10. Dierks, T.: The transport layer security (TLS) protocol version 1.2. RFC 5246 (2008)
11. Dierks, T., Rescorla, E.: The transport layer security (TLS) protocol version 1.3. RFC 5246bis (2014)
12. Dingledine, R., Mathewson, N., Syverson, P.: Tor: The second-generation onion router. Tech. rep., DTIC Document (2004)
13. Duong, T., Rizzo, J.: BEAST—here come the XOR Ninjas. http://nerdoholic.org/uploads/dergln/beast_part2/ssl_jun21.pdf (2011)
14. Electronic Frontier Foundation: Https everywhere. https://www.eff.org/https-everywhere (2013)
15. Ettercap Project: Ettercap home page. http://ettercap.github.io/ettercap/ (2013)
16. Evans, C., Palmer, C., Sleevi, R.: Public key pinning extension for HTTP. IETF Internet Draft (2014)
17. Farrell, S., Hoffman, P., Thomas, M.: HTTP Origin-Bound Authentication (HOBA). IETF Internet Draft (2014)
18. Franks, J., Hallam-Baker, P., Hostetler, J., Lawrence, S., Leach, P., Luotonen, A., Stewart, L.: HTTP authentication: basic and digest access authentication. RFC Draft Standard (RFC 2617) (1999)
19. Friedl, S., Popov, A.: Transport Layer Security (TLS) application layer protocol negotiation extension. RFC Proposed Standard (RFC 7301) (2014)
20. Gluck, Y., Harris, N., Prado, A.: BREACH: reviving the cRIME attack. http://breachattack.com/resources/BREACH%20-%20SSL,%20gone%20in%2030%20seconds.pdf (2013)
21. Goland, Y., Whitehead, E., Faizi, A., Carter, S., Jensen, D.: HTTP extensions for distributed authoring—WEBDAV (1999)
22. Grant, A.C.: Search for trust: an analysis and comparison of CA system alternatives and enhancements (2012)

23. HAK5: wifi pineapple. https://wifipineapple.com/ (2013)
24. Hodges, J., Jackson, C., Barth, A.: HTTP strict transport security (HSTS). RFC Proposed Standard (RFC 6797) (2012)
25. Hoffman, P., Schlyter, J.: The DNS-based authentication of named entities (DANE) transport layer security (TLS) protocol: TLSA. RFC Proposed Standard (RFC 6698) (2012)
26. Huang, L.S., Rice, A., Ellingsen, E., Jackson, C.: Analyzing forged ssl certificates in the wild. In: Proceedings of the 35th IEEE Symposium on Security and Privacy (SP) (2014)
27. Jackson, C., Barth, A.: ForceHTTPS: protecting high-security web sites from network attacks. In: Proceedings of the 17th International Conference on World Wide Web (WWW), pp. 525–534 (2008)
28. Langley, A.: Overclocking ssl. https://www.imperialviolet.org/2010/06/25/overclocking-ssl.html (2010)
29. Langley, A.: ChaCha20 and Poly1305 based Cipher suites for TLS. IETF Internet Draft (2013)
30. Laurie, B., Langley, A., Kasper, E.: Certificate transparency. RFC Experimental (RFC 6962) (2013)
31. Lennon, M.: Hackers exploited heartbleed bug to steal 4.5 million patient records: Report. http://www.securityweek.com/hackers-exploited-heartbleed-bug-steal-45-million-patient-records-report (2014)
32. Marlinspike, M.: New tricks for defeating ssl in practice. BlackHat DC, February (2009)
33. Marlinspike, M.: Sslstrip. http://www.thoughtcrime.org/software/sslstrip/ (2009)
34. Masnick, M.: FLYING PIG: The NSA is running man in the middle attacks imitating Google's servers. http://www.techdirt.com/articles/20130910/10470024468/flying-pig-nsa-is-running-man-middle-attacks-imitating-googles-servers.shtml (2013)
35. Modell, M., Barz, A., Toth, G., Loesch, C.v.: Certificate patrol. https://addons.mozilla.org/en-US/firefox/addon/certificate-patrol/ (2014)
36. Nikiforakis, N., Younan, Y., Joosen, W.: Hproxy: client-side detection of ssl stripping attacks. In: Proceedings of the 7th Conference on Detection of Intrusions and Malware, and Vulnerability Assessment (DIMVA), pp. 200–218 (2010)
37. Nottingham, M.: Opportunistic encryption for HTTP URIs. IETF Internet Draft (2014)
38. Prins, J.: Diginotar certificate authority breach—'operation black tulip'. Fox-IT (2011)
39. Qualys: Qualys SSL labs. https://www.ssllabs.com/ (2014)
40. Qualys: Trustworthy internet movement—ssl pulse. https://www.trustworthyinternet.org/ssl-pulse/ (2014)
41. Rescorla, E., Ray, M., Dispensa, S., Oskov, N.: Transport layer security (TLS) renegotiation indication extension. RFC Proposed Standard (RFC 5746) (2010)
42. Ristić, I.: OpenSSL cookbook. Feisty Duck (2013)
43. Ristić, I.: Bulletproof SSL and TLS. Feisty Duck (2014)
44. Rizzo, J., Duong, T.: The CRIME Attack. https://docs.google.com/presentation/d/11eBmGiHb YcHR9gL5nDyZChu_-lCa2GizeuOfaLU2HOU/edit?pli=1#slide=id.g1d134dff_1_222(2012)
45. Roberts, P.: Infographic: A heartbleed disclosure timeline (secunia). https://securityledger.com/2014/06/infographic-a-heartbleed-disclosure-timeline-secunia/ (2014)
46. Schneier, B.: Hearbleed. https://www.schneier.com/blog/archives/2014/04/heartbleed.html (2014)
47. Schoen, S., Galperin, E.: Iranian man-in-the-middle attack against google demonstrates dangerous weakness of certificate authorities. https://www.eff.org/deeplinks/2011/08/iranian-man-middle-attack-against-google (2011)
48. Sheffer, Y., Holz, R., Saint-Andre, P.: Recommendations for secure use of TLS and DTLS. IETF Internet Draft (2014)
49. Song, D.: dsniff. http://www.monkey.org/ dugsong/dsniff/ (2000)
50. The Guardian: Edward Snowden. http://www.theguardian.com/world/edward-snowden (2013)
51. The H Security: trustwave issued a man-in-the-middle certificate. http://h-online.com/-1429982 (2012)

52. Toussain, M., Shields, C.: Subterfuge. http://kinozoa.com/blog/subterfuge-documentation/ (2013)
53. W3Techs: Usage statistics and makert share of ssl certificate authorities for websites, august 2014. http://w3techs.com/technologies/overview/ssl_certificate/all (2014)
54. Wi-Fi Alliance: Wi-Fi protected access: strong, standards-based, interoperable security for today's Wi-Fi networks. http://www.ans-vb.com/Docs/Whitepaper_Wi-Fi_Security4-29-03.pdf (2003)

Chapter 6
Attacks on the Browser's Requests

The previous chapter covered network-level attacks, allowing the attacker to listen to information, or modify traffic being sent. The attacks in this chapter target the browser's requests, and the attacker actually runs code within the victim's browser, instead of sitting remotely on the network.

By attacking the browser's requests, an attacker is able to forge requests to a target application in the user's name. The attacker tricks the user's browser into sending a forged request, generally without the user noticing the request being sent to the target application. The core problem behind forging requests is the fact that a target application often cannot distinguish between legitimate requests, made by the user, and forged requests, made without the user's consent. Due to the way the Web platform works, it is impossible to determine whether a request is legitimate without taking some additional measures.

In this chapter, we cover two important ways of forging requests in the user's name. The first is *cross-site request forgery* (CSRF), where the attacker tricks the user's browser into automatically sending requests to the target application. CSRF attacks can cause several actions, such as updating profile information or actually carrying out transactions, such as wire transfers in online banking software. The second way is *UI redressing*, where the attacker tricks the user into interacting with a seemingly innocent page, while the interactions are actually sent to the target application. UI redressing attacks can cause users to unknowingly post status updates on social networks or enable their webcam in the settings of the Flash player.

6.1 Cross-Site Request Forgery

A CSRF attack enables an attacker to forge requests to the target application from a legitimate user's browser. A vulnerable application handles these forged requests the same way as legitimate requests from the victim. Successful CSRF attacks can trigger many actions in vulnerable applications, such as modifying account settings or stealing money through an online banking system [39].

© Philippe De Ryck, Lieven Desmet, Frank Piessens, Martin Johns 2014
P. De Ryck et al., *Primer on Client-Side Web Security*,
SpringerBriefs in Computer Science, DOI 10.1007/978-3-319-12226-7_6

The CSRF is prevalent in modern Web sites, and is ranked in both the OWASP Top 10 project [37] and the CWE/SANS Top 25 Most Dangerous Programming Errors [25]. Both small and large-scale projects are affected with, for example, CSRF vulnerabilities in online banking systems [39], Gmail [14], and eBay [18].

6.1.1 Description

The goal of a CSRF attack is to forge a request from a victim's browser to the target application, triggering state-changing effects in the target application. Examples of such state-changing effects are modifying account settings or adding items to a shopping cart. For a CSRF attack to succeed, it is essential that the user is already authenticated to the target application, since the user will not see the forged request. From the application's point of view, the forged request has the same structure as a legitimate request, and is therefore indistinguishable from a legitimate request.

Tricking the victim's browser into making a request to the target application is a straightforward task. Browsers not only frequently issue requests to numerous unrelated sites, for instance when loading external resources such as images, style sheets, or a document to load in a frame, but also when submitting form data to a cross-origin URI. An attacker can easily include code that triggers a request to the target application in his own Web site or a site he controls. Alternatively, he can inject HTML or JavaScript code in an unrelated but legitimate site, such as a Web forum allowing users to post images or other content. These two attack vectors require the capabilities of a Web attacker and forum poster. Listing 6.1 shows the code that uses a hidden image to trigger a forged request and Listing 6.2 shows a cross-origin form submission.

```
<img width="0" height="0"
          src="http://admin.example.com/deleteAccount.php" />
```

Listing 6.1 A CSRF attack carried out by a hidden *img* tag that triggers a GET request to the target application.

```
document.getElementById("somediv").innerHTML += "<iframe
          id='attackframe' style='height: 0px; width: 0px;'></iframe>";
var f = document.getElementById("attackframe");

var code =   "<form id='attackform' action='http://admin.example.com/
          createAccount.php' method='POST'>";
code +=      "<input type='hidden' name='username' value='attacker'>";
code +=      "<input type='hidden' name='password' value='12345678'>";
code +=      "<input type='hidden' name='action' value='create'>";
code +=   "</form>";

f.contentDocument.body.innerHTML = code;
f.contentDocument.getElementById("attackform").submit();
```

Listing 6.2 A CSRF attack carried out by JavaScript code that creates a hidden *iframe* containing a *form*, which is then automatically submitted to the target application.

Fig. 6.1 In the CSRF attack depicted here, the attacker triggers a request from origin E to origin A (step 13), to which the browser attaches the cookies from the existing session with origin A. If origin A does not have CSRF protection, this request will be executed as if it was generated by the user

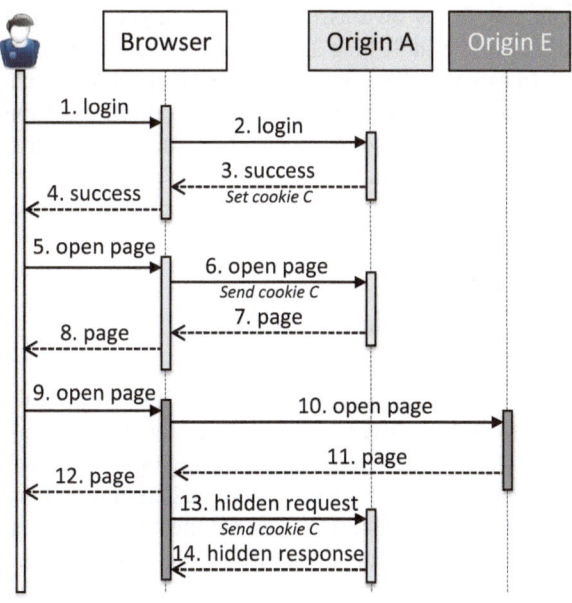

A CSRF attack can only be successful if the forged request happens within a previously authenticated session between the victim's browser and the target application. Unfortunately, the design of current session management mechanisms in the browser attaches session information to any outgoing request, fostering the prevalence of CSRF attacks. For example, the browser attaches the relevant cookies for the domain, scheme, and path to each outgoing request, both for requests internal to the application and cross-site or cross-application requests (illustrated in Fig. 6.1). Additionally, many applications prefer long-life sessions, sticking around as long as the browser remains open, regardless of whether an application is currently active in a browser tab. Essentially, this means that if a user had an authenticated session with the target application in the lifetime of the browser, it is likely that forged requests within an authenticated session can be made.

A *Login CSRF* attack is a variation of a CSRF attack, where the attacker forges a request to authenticate the victim with an attacker-chosen account. Essentially, the attacker submits a login form from within the user's browser, using the credentials of the attacker. When the user unknowingly uses the targeted application, any entered information is associated with the attacker-chosen account, and can potentially leak to the attacker. A common example is a search engine keeping a history for authenticated users.

Next to traditional login CSRF attacks that forge a submission of the target application's authentication form, login CSRF attacks can also target applications using third-party authentication providers such as Google single sign-on, Facebook authentication, or OpenID. These authentication providers provide the target application with an *assertion*, containing the necessary information to confirm a successful authentication, as well as the user's identity. Using a login CSRF attack, an attacker can submit his own assertion to the target application from the victim's browser, effectively establishing an authenticated session tied to the attacker's credentials.

In essence, the problem of a CSRF attack is the lack of intent, leaving the server in the dark as to whether a request was made intentionally by legitimate application code, or was forged by an attacker. The fact that browsers handle same-origin and cross-origin requests identically, and Web applications now heavily depend on this behavior, enables CSRF attacks and hampers effective countermeasures.

6.1.2 Mitigation Techniques

During the early years of CSRF, several simple mitigation techniques have been proposed, but proven ineffective at protecting against CSRF attacks. One suggestion is to only carry out state-changing operations using POST requests, as actually mandated by HTTP specification [12], assuming that forging POST requests is not feasible. Unfortunately, this is not the case [39], as shown by the code example in Listing 6.2, rendering this advice useless in protecting against CSRF.

A second mitigation technique enforces referrer checking at the server side. State-changing requests should only be accepted if the value of the `Referer` header[1] contains a trusted site and rejected otherwise. Referrer checking would effectively mitigate CSRF attacks, were it not that the presence of the `Referer` header in the request headers is unreliable. The `Referer` header is often missing due to privacy concerns, since it tells the target application which resource at which URI triggered the request. Similarly, browsers do not add the header when an HTTPS resource, which is considered sensitive, refers to an HTTP resource. Additionally, browser settings, corporate proxies, privacy proxies, or extensions [28, 29] and referrer-anonymizing services [26] enable the stripping of automatically added `Referer` headers.

As an improvement to the `Referer` header, the `Origin` header provides the server with information about the origin of a request, without the strong privacy-invasive nature of the `Referer` header [3]. Unfortunately, the specification [2] only states that the `Origin` header *may* be added, but does not require user agents to do so, potentially causing the same problems as with the `Referer` header. The `Origin` header, however, is mandatory when using cross-origin resource sharing (CORS) [36], an API that enables the sharing of resources across origins.

Alternatively, token-based approaches are an effective countermeasure against CSRF attacks [6]. A token-based approach adds a unique token to the code triggering state-changing operations. When the browser submits the request leading to the action, the token is included automatically, and verified by the server. Token-based approaches prevent the attacker from including a valid token in his payload, causing

---

[1] The `Referer` header was originally misspelled in the specification, and the header has kept this name until this day. In text, the correctly-spelled *referrer* is more commonly used.

the request to be rejected. Key to the success of this mitigation technique is keeping the token for an action out of the attacker's reach. This requires the tokens to be unique, or at least bound to a specific user. Additionally, the tokens are embedded in the page, where they are protected by the same-origin policy, preventing theft by an attacker-controlled context, loaded in the same browser. One example of a token-based approach are hidden form fields that contain a randomly generated, user-bound token, which is submitted with the form's contents but cannot be read from the DOM by a browsing context from another origin.

Further research on token-based approaches, which often struggle with Web 2.0 applications, has yielded several improvements over traditional tokens, to enable complex client-side scripting and cross-origin requests between cooperating sites. jCSRF [27], a server-side proxy solution, transparently adds security tokens to client-side resources and verifies the validity of incoming requests. Alternatively, double-submit techniques [20] embed a nonce in two different locations, for example, in a cookie and as a hidden form field, allowing the server to compare both values, without keeping track of state. Since the attacker cannot manipulate both tokens, he is unable to forge valid requests.

Another approach at the level of the application's architecture is based on the observation that the cross-origin accessibility of Web application resources allows the attacker to target any resource by making a request from a different origin. Several techniques propose to mitigate CSRF by fixing the set of entry points to known safe resources, thus eliminating a CSRF attack on a sensitive resource. These entry points can be enforced purely at the server side [7] or in combination with a browser-based mechanism [8]. Recently, the concept of entry points gained traction with browser vendors and is being integrated as a core feature [30].

Another effective mitigation of CSRF attacks involves explicit user approval of state-changing operations. By requiring additional, unforgeable user interaction, the attacker is unable to complete the CSRF attack in the background. Examples are explicit reauthentication for sensitive operations or the use of an out-of-band device to generate security tokens, as employed by many European online banking systems. The risk associated with this mitigation technique is a shift in attack from CSRF to clickjacking, which is covered in Sect. 6.2.

Finally, client-side solutions have emerged to protect legacy applications, which are no longer updated, or where developers do not know or care about CSRF vulnerabilities, leaving users vulnerable in the end. These client-side solutions detect potentially dangerous requests and either block them or strip them from implicit authentication credentials, such as cookies. Examples are RequestRodeo [17], the first client-side mitigation technique in the form of a proxy, followed by browser extensions CsFire [9, 10], RequestPolicy [34], DeRef [13], and NoScript ABE [23]. While these client-side solutions have registered some success among enthusiasts, their main disadvantage is the need for compatibility with all sites, often resulting in false positives, which distort the delicate balance between security and usability.

6.1.3 State of Practice

Current practices for mitigating CSRF attacks are focused on token-based approaches, either custom-built for the application or deployed as part of a Web framework, such as Ruby on Rails, CodeIgniter, and several others. Alternatively, server-side libraries or APIs offer CSRF protection as well, such as the community-supported OWASP ESAPI API and CSRFGuard. Sites being built using a content management system (CMS)—instead of being built from scratch—can benefit from built-in CSRF support as well. For example, Drupal, Django, and WordPress offer token-based CSRF protection, with Drupal even extending its support to optional customized modules.

Applications using the OAuth protocol for authentication are vulnerable to login CSRF attacks, as shown by a formal analysis of Web site authentication flows [1], which has dubbed this problem as *social login CSRF*. OAuth is a protocol enabling third-party clients limited access to an API, such as used in Facebook authentication [11]. The OAuth specification recommends using a generated nonce, strongly bound to the user's session, which would prevent a social login CSRF attack, if followed by the implementations of the protocol.

6.1.4 Best Practices

Ideally, Web developers mitigate CSRF attacks by using built-in protection mechanisms for state-changing operations. Alternatively, custom token-based approaches can be integrated as well, taking precautions to prevent token compromise. For legacy applications, the use of a transparent server-side solution, such as CSRFGuard, can enable CSRF protection without having to fiddle with the application code.

As a second line of defense, it is recommended that user involvement should be required for truly sensitive operations, especially when they have direct financial consequences, or can lead to the compromise of a user account. For example, it is not unreasonable to require explicit user involvement or reauthentication when changing the password or making a wire transfer in an online banking system.

6.2 UI Redressing

A UI redressing attack, also known as clickjacking or tapjacking, redresses or "redecorates" a target application, confusing the user who is interacting with the application. For example, Fig. 6.2 illustrates a clickjacking attack using a transparent overlay. UI redressing attacks can be used to trigger any user interaction within the target application, such as clicking a button, dragging and dropping items, etc.

A UI redressing attack uses various innocent features, combining them to trick the user into clicking a sensitive element. UI redressing attacks can not only be annoying, but also malicious. Examples of the former are Tweetbombs [21], which post Twitter

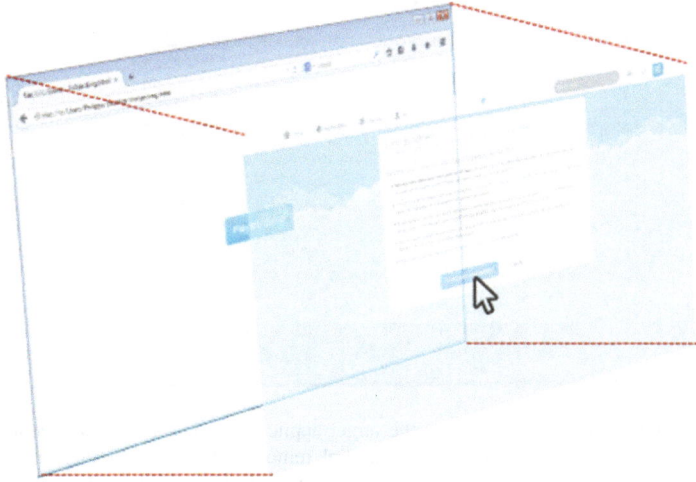

Fig. 6.2 The essence of a clickjacking attack is tricking the user into clicking on a specific location, under which an element of the target application is positioned. In this example, the user thinks he starts a game, but in fact clicks on a button in the hidden target application

status updates to the victim's account, and LikeJacking, which triggers unintended likes on Facebook pages. Examples of the latter are attacks that trick the user into enabling webcam access for the Flash player [16], and attacks on wireless routers, stealing the secret WPA keys [33].

6.2.1   Description

The goal of a UI redressing attack is to forge a request from the victim's browser to the target application, by making the user unintentionally interact with an element on a page of the target application. An attacker achieves this using misdirection, by redressing the UI of the page to hide the real element that will be clicked by the user. Many forms of UI redressing are possible, from transparent overlays to very precise positioning of elements, or even fake cursors stealing the user's attention.

A UI redressing attack requires a coordinating application, which is under control of the attacker and actually attracts legitimate interaction from the user. However, the coordinating application masquerades the target application, causing the user's interaction to be directed towards the target application. In the example in Fig. 6.2, the user actually thinks he clicks on the *Play!* button, but in reality, the click goes towards the invisibly framed page. Since both buttons are precisely positioned on top of each other, the attacker ensures that the user actually clicks at the right location.

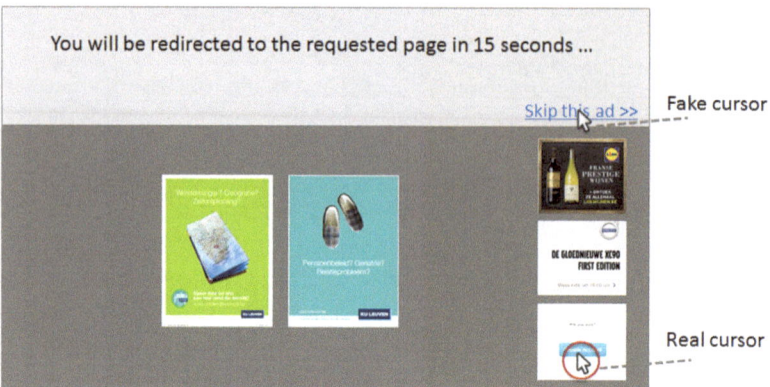

Fig. 6.3 A cursorjacking attack, where the target application's *Delete Account* button is at the bottom right of the page, with a *skip this ad* bait link remotely above it. Note there are two cursors displayed on the page: a fake cursor is drawn over the skip this ad link while the actual pointer hovers over the *delete account* button

Note that the attacker requires the capabilities of a Web attacker but does not control the target application, and that the interactions of the user with the target application are indistinguishable from legitimate interactions. Mitigation techniques for CSRF attacks are ineffective against UI redressing attacks, since the requests are not cross-origin but actually originate from within the application.

While UI redressing attacks were traditionally known as clickjacking attacks, numerous variations have emerged as the technology evolved. *Double clickjacking* [15] tricks the user into double clicking, and quickly raises a previously opened pop-under window after the first click, misdirecting the user's click to the target application, which is opened in the pop-under window. Another variation uses *history navigation* to store a target page in the window's history, and quickly switching back when the user double clicks somewhere in the attacker's page [38]. Instead of abusing clicking, an attack can also use the features of the drag-and-drop API [5] to persuade the user to drag some text into the target application, thereby injecting data into form fields [35]. Another variation is a *cursorjacking* attack [16], where a fake cursor is drawn on the screen, while the original cursor is hidden or out of the user's focus on screen (Fig. 6.3). Alternatively, *strokejacking* abuses keyboard focus features to trick the user into typing in an input field of the target application [16]. Finally, *tapjacking* brings UI redressing attacks to mobile devices, tricking the user into tapping on hidden elements [33].

In essence, a UI redressing attack results in an unintentional request by misdirecting the interaction of a user. Well-conducted UI redressing attacks are impossible to observe, exonerating the users from all blame. UI redressing attacks can even be used to bypass an application's mitigation techniques, such as an explicit confirmation request before performing sensitive actions.

6.2.2   Mitigation Techniques

UI redressing attacks commonly use frames to embed the target application. Traditional mitigation techniques therefore use *framebusting code*. Framebusting code is targeted at detecting the unpermitted framing of an application, and subsequently breaking out of the frame by moving the application to a top-level frame, as shown by the example in Listing 6.3). Simple framebusting code is often easily evaded, but carefully constructing robust framebusting code can withstand evasion, or fail in safe ways [32]. The downside of framebusting is the strict on or off mode, either allowing all kinds of framing or no framing at all, not even by trusted applications. In the modern Web, with mashups and composed applications, this might be problematic.

A second popular mitigation technique is the regulation of framing through the X-Frame-Options header [31]. By adding this header to the response, an application can indicate that framing is denied, allowed within its origin, or allowed by an explicitly listed origin. Recently, the functionality offered by the X-Frame-Options header has been integrated in the content security policy with the *frame-ancestors* directive [4]. The *frame-ancestors* directive offers better support for nested browsing contexts, which are still vulnerable with the X-Frame-Options header [19], and supports multiple host source values, instead of the one supported by the X-Frame-Options header.

Research on UI redressing attacks has also focused on a browser-supported solution that addresses the root cause, i.e., user misdirection. *InContext* [16] incorporates several measures that ensure that a user's click is genuine, for example, by comparing screenshots at the time of the click or by highlighting the area of the cursor to prevent attacks involving a fake cursor. *InContext* also serves as an inspiration for the new standardization efforts by W3C to ensure UI integrity in Web applications, effectively preventing UI redressing attacks [24]. The specification proposes several new directives to include in the content security policy, giving the developer control over several heuristics to determine the genuineness of the interaction. Similar to other directives in the Content Security Policy, the browser will enforce these heuristics, by blocking and reporting any violation.

```
//UNSAFE - DO NOT USE
if(top != self) top.location.replace(location);
```

Listing 6.3 A simple approach aimed at detecting unpermitted framing, and breaking out of the frame by moving the application to the top-level frame. While this countermeasure is often used, it is easily evaded [32]

Finally, clickjacking can be combated from the client side as well. The popular security add-on NoScript [22] includes the ClearClick module, which does a screenshot-based comparison of the area to be clicked with the actually clicked element, and which served as an inspiration for the *InContext* work [16]. When a difference between both is detected, the user is warned and explicitly asked to confirm the action before the request is sent.

6.2.3 State of Practice

By design, all Web applications are vulnerable to clickjacking attacks, but the attack receives little attention compared to higher-risk attacks. Users of major, well-known Web applications such as Twitter, Facebook, etc. have fallen victim to clickjacking attacks, often indicated by spam messages making their way through the application.

Many Web applications deploy some form of framebusting code, of which several variations are known to be vulnerable to evasion [32]. Additionally, many applications have a different front end for normal browsers and mobile browsers, often only implementing framebusting in their normal version [33]. Currently, the X-Frame-Options header is gaining adoption. Our July 2014 survey of the Alexa top 10,000 domains discovered 2159 domains that include an X-Frame-Options header in their responses.

6.2.4 Best Practices

Ideally, applications employ both effective framebusting code where possible (illustrated in Listing 6.4), combined with a framing restriction, either with the X-Frame-Options header or through CSP's *frame-ancestors* directive, configured as tightly as possible. Introducing additional user interactions, such as an explicit confirmation dialog, will certainly make UI redressing attacks more difficult but will not always be sufficient to eradicate them [16].

In the near future, the newly standardized User Interface Safety Directives for CSP will become available [24], giving more fine-grained control to determine whether the interaction is genuine.

```
if( self == top ) {
  document.documentElement.style.display = 'block' ;
} else {
  top.location = self.location ;
}
```

Listing 6.4 Combining this framebusting code with content that is hidden by default offers clickjacking protection that cannot be evaded, and it fails when the detection is hampered somehow.

References

1. Bansal, C., Bhargavan, K., Maffeis, S.: Discovering concrete attacks on website authorization by formal analysis. In: Proceedings of the 25th IEEE Computer Security Foundations Symposium (CSF), pp. 247–262 (2012)
2. Barth, A.: The web origin concept. RFC 6454 (2011)

3. Barth, A., Jackson, C., Mitchell, J.C.: Robust defenses for cross-site request forgery. In: Proceedings of the 15th ACM Conference on Computer and Communications Security (CCS), pp. 75–88 (2008)
4. Barth, A., Veditz, D., West, M.: Content security policy level 2. W3C Working Draft (2014)
5. Berjon, R., Faulkner, S., Leithead, T., Navara, E.D., O'Connor, E., Pfeiffer, S., Hickson, I.: HTML 5.1 specification. W3C Working Draft (2014)
6. Burns, J.: Cross site reference forgery: An introduction to a common Web application weakness. https://www.isecpartners.com/media/11961/csrf_paper.pdf (2005)
7. Chen, E.Y., Bau, J., Reis, C., Barth, A., Jackson, C.: App isolation: get the security of multiple browsers with just one. In: Proceedings of the 18th ACM Conference on Computer and Communications Security (CCS), pp. 227–238 (2011)
8. Czeskis, A., Moshchuk, A., Kohno, T., Wang, H.J.: Lightweight server support for browser-based CSRF protection. In: Proceedings of the 22nd International Conference on World Wide Web (WWW), pp. 273–284 (2013)
9. De Ryck, P., Desmet, L., Heyman, T., Piessens, F., Joosen, W.: CsFire: Transparent client-side mitigation of malicious cross-domain requests. In: Proceedings of the 2nd International Symposium on Engineering Secure Software and Systems (ESSoS), pp. 18–34 (2010)
10. De Ryck, P., Desmet, L., Joosen, W., Piessens, F.: Automatic and precise client-side protection against csrf attacks. In: Proceedings of the 16th European Symposium on Research in Computer Security (ESORICS), pp. 100–116 (2011)
11. Facebook: Facebook login. http://developers.facebook.com/docs/facebook-login/ (2013)
12. Fielding, R., Gettys, J., Mogul, J., Frystyk, H., Masinter, L., Leach, P., Berners-Lee, T.: Hypertext Transfer Protocol—HTTP/1.1. RFC 2616 (1999)
13. Fung, B.S., Lee, P.P.: A privacy-preserving defense mechanism against request forgery attacks. In: Proceedings of the 10th International Conference on Trust, Security and Privacy in Computing and Communications (TrustCom), pp. 45–52 (2011)
14. Hepper, D.: Gmail CSRF vulnerability explained. http://daniel.hepper.net/blog/2008/11/gmail-csrf-vulnerability-explained/ (2008)
15. Huang, L.S., Jackson, C.: Clickjacking attacks unresolved. https://docs.google.com/document/pub?id=1hVcxPeCidZrM5acFH9ZoTYzg1D0VjkG3BDW_oUdn5qc (2011)
16. Huang, L.S., Moshchuk, A., Wang, H.J., Schechter, S., Jackson, C.: Clickjacking: attacks and defenses. In: Proceedings of the 21st USENIX Security Symposium, pp. 22–22 (2012)
17. Johns, M., Winter, J.: Requestrodeo: Client side protection against session riding. In: Proceedings of the OWASP AppSec Europe 2006 Conference (AppSecEU), pp. 5–17 (2006)
18. Kovacs, E.: CSRF Vulnerability in eBay allows hackers to hijack user accounts. http://news.softpedia.com/news/CSRF-Vulnerability-in-eBay-Allows-Hackers-to-Hijack-User-Accounts-Video-383316.shtml (2013)
19. Lekies, S., Heiderich, M., Appelt, D., Holz, T., Johns, M.: On the fragility and limitations of current browser-provided clickjacking protection schemes. In: Proceedings of the 6th USENIX Workshop on Offensive technologies (WOOT), pp. 53–63 (2012)
20. Lekies, S., Tighzert, W., Johns, M.: Towards stateless, client-side driven cross-site request forgery protection for Web applications. In: Proceedings of the 7th conference on Sicherheit, Schutz und Zuverlässigkeit (Sicherheit), pp. 111–121 (2012)
21. Mahemoff, M.: Explaining the dont click clickjacking tweetbomb. http://softwareas.com/explaining-the-dont-click-clickjacking-tweetbomb/ (2009)
22. Maone, G.: NoScript—JavaScript/Java/Flash blocker for a safer Firefox experience! http://noscript.net/ (2013)
23. Maone, G.: NoScript Application Boundaries Enforcer (ABE). http://noscript.net/abe/ (2013)
24. Maone, G., Huang, D.L.S., Gondrom, T., Hill, B.: User interface safety directives for content security policy. W3C Last Call Working Draft (2014)
25. Martin, B., Brown, M., Paller, A., Kirby, D.: Cwe/sans top 25 most dangerous programming errors. http://cwe.mitre.org/top25/ (2011)

26. Nikiforakis, N., Van Acker, S., Piessens, F., Joosen, W.: Exploring the ecosystem of referrer-anonymizing services. In: Proceedings of the 12th Privacy Enhancing Technologies Symposium (PETS), pp. 259–278 (2012)
27. Pelizzi, R., Sekar, R.: A server-and browser-transparent csrf defense for web 2.0 applications. In: Proceedings of the 27th Annual Computer Security Applications Conference (ACSAC), pp. 257–266 (2011)
28. Privoxy. Online at http://www.privoxy.org (2013)
29. RefControl. https://addons.mozilla.org/en-us/firefox/addon/refcontrol/ (2013)
30. Ross, D.: Entry point regulation for web apps. http://randomdross.blogspot.be/2014/08/entry--point-regulation-for-web-apps.html (2014)
31. Ross, D., Gondrom, T.: HTTP header field X-frame-options. RFC Informational (RFC 7034) (2013)
32. Rydstedt, G., Bursztein, E., Boneh, D., Jackson, C.: Busting frame busting: a study of clickjacking vulnerabilities at popular sites. Web 2.0 Security and Privacy (W2SP) (2010)
33. Rydstedt, G., Gourdin, B., Bursztein, E., Boneh, D.: Framing attacks on smart phones and dumb routers: tap-jacking and geo-localization attacks. In: Proceedings of the 4th USENIX Workshop on Offensive technologies (WOOT), pp. 1–8 (2010)
34. Samuel, J., Zhang, B.: Requestpolicy: Increasing web browsing privacy through control of cross-site requests. In: Proceedings of the 9th Privacy Enhancing Technologies Symposium (PETS), pp. 128–142 (2009)
35. Stone, P.: Next generation clickjacking. BlackHat Europe (2010)
36. van Kesteren, A.: Cross-origin resource sharing. W3C Recommendation (2014)
37. Wichers, D.: Owasp top 10. https://www.owasp.org/index.php/Category:OWASP_Top_Ten_Project (2013)
38. Zalewski, M.: Arbitrary page mashups (ui redressing). http://code.google.com/p/browsersec/wiki/Part2#Arbitrary_page_mashups_(UI_redressing) (2010)
39. Zeller, W., Felten, E.W.: Cross-site request forgeries: exploitation and prevention. Tech. rep., Princeton University (2008)

Chapter 7
Attacks on the User's Session

The previous chapter covered attacks on the user's requests, enabling an attacker to send requests from within the browser. Attacks on the user's session, covered in this chapter, generally have a higher impact, as they give the attacker full control over the user's session.

By gaining control over an authenticated session, the attacker gets the same level of access to the target application as the victim. Some attacks allow the attacker to obtain the user's authenticated session in his own browser, while other attacks focus on using a user's stolen credentials. These attacks are enabled by applications deploying weak authentication systems and insufficiently protecting authenticated sessions.

This chapter covers two ways of transferring an existing authenticated session from the victim's browser to the attacker's browser: *session hijacking* and *session fixation*. Finally, the *use of stolen credentials*, which gives the attacker all he needs to independently establish an authenticated session is discussed.

7.1 Session Hijacking

A session hijacking attack allows the attacker to transfer an authenticated session from the victim's browser to an attacker-controlled browser. Using the transferred session, the attacker can impersonate the user and perform all actions available to the user.

Session hijacking, together with other session-related problems, is ranked second in the Open Web Application Security Project (OWASP) top ten [51]. In addition, with ubiquitous, freely accessible, and unprotected wireless networks, session hijacking has become a straightforward attack [23].

7.1.1 Description

The goal of a session hijacking attack is to transfer the user's authenticated session to a different machine or browser, enabling the attacker to continue working in the victim's session. To achieve this, the attacker hijacks the session, that the user has

© Philippe De Ryck, Lieven Desmet, Frank Piessens, Martin Johns 2014   69
P. De Ryck et al., *Primer on Client-Side Web Security,*
SpringerBriefs in Computer Science, DOI 10.1007/978-3-319-12226-7_7

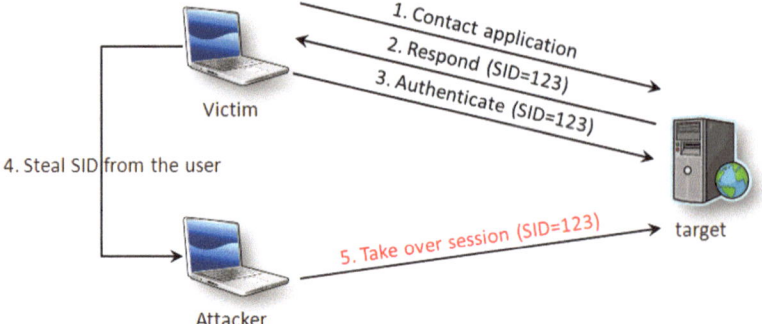

Fig. 7.1 In a *session hijacking* attack, an attacker steals the session identifier of the user (step 4), resulting in a complete compromise of the user's session

established with the target application. Note that if the attacker manages to hijack an unauthenticated session, he simply has to wait until the user authenticates himself, since this state will be stored in the server-side session object.

Technically, once a session between the user's browser and the server is established, future requests will be handled within the context of this session (explained in Chap. 3). The de facto standard session management mechanism in modern Web applications is cookie-based, where a random, unique session identifier is stored in a cookie within the browser. In a session hijacking attack, an attacker succeeds in stealing the session identifier, which he can subsequently use to send requests to the server (illustrated in Fig. 7.1). This is possible because the session identifier acts as a *bearer token*, and the mere presence of this identifier in a cookie attached to the request suffices for legitimizing the request within the session.

Depending on the security parameters of the cookie, an attacker has several ways of obtaining the session identifier. One way is by calling the *document.cookie* property from within the target's application origin, which can for example be achieved through cross-site scripting (covered in Chap. 8). A second way is by directly accessing the cookie store from a compromised browser, for example, by installing a malicious browser extension (covered in Chap. 9). A third alternative is by eavesdropping on the network traffic (covered in Chap. 5) and snatching the session identifier from the response or any subsequent request, as illustrated by point-and-click tools such as Firesheep [9]. Finally, a weak or predictable session identifier can be guessed or obtained through a brute-force attack.

An alternative session management mechanism is based on the uniform resource identifier (URI) parameters, including a session identifier as a parameter in the URI in every request to the application. This mechanism is often used as a fallback mechanism for browsers that do not support cookies or refuse to store them. Technically, the scenario for a session hijacking attack does not change, other than the means to obtain the session identifier. An attacker can still access it from JavaScript or eavesdrop on the network to extract it. In addition, an attacker can attempt to trigger a request to an attacker-controlled resource, hoping that a `Referer` header will be included, since it contains the full URI, including the parameter with the session identifier.

In essence, a session hijacking attack is possible because the session identifier, which acts as a bearer token for an authenticated session, is easily obtained and transferable between browsers. Making the session identifier accessible through JavaScript or by eavesdropping on the network is a suboptimal decision, which enables a highly dangerous and harmful attack.

7.1.2   Mitigation Techniques

A traditional mitigation technique for session hijacking is *IP address binding*, where the server binds a user's session to a specific IP address. Subsequent requests within this session need to come from the same IP address, and any requests coming from another IP address are discarded. While this mitigation technique works well in scenarios where every machine has a unique, unchanging public IP address, it is ineffective when the same public IP address is shared among multiple machines, or when the public IP address changes during a session. Precisely, these two cases have become ubiquitous in modern network infrastructure, with NATed home and company networks (publicly accessible), shared wireless networks, and mobile networks. Recently, the technique of tracking a client has been refined through browser fingerprinting, where numerous characteristics of the browser are compiled into a fingerprint [1, 20, 40]. Anomaly detection based on the browser fingerprint triggers alerts when an unexpected fingerprint is seen, which may be an attacker stealing a session.

Another approach focuses on preventing the theft of the session identifier, which is commonly stored and transmitted in a cookie. The *HttpOnly* and *Secure* cookie attributes can be used to, respectively, prevent a cookie from being accessible through JavaScript and prevent a cookie issued over HTTPS from being used (and thus leaked) on a non-HTTPS connection. Correctly applying both attributes to cookies holding a session identifier effectively thwarts script-based session hijacking attacks, as well as session hijacking attacks through eavesdropping on network traffic.

One long-lived line of research focuses on providing protection against session hijacking attacks from within a Web application, without specific infrastructure support at the client side. The idea behind these approaches is not to hide the session identifier but to ensure that the session identifier no longer acts as a bearer token, meaning that the mere knowledge of the session identifier is insufficient to hijack a session.

SessionSafe [31] combines several mitigation techniques against session hijacking into a single countermeasure, and thereby effectively prevents script-based session hijacking attacks. To summarize, three combined mitigation techniques are (i) deferred loading, which hides the session identifier from malicious JavaScript before main content is loaded, (ii) one-time URLs, where a secret component prevents URLs from being guessed by an attacking script, and (iii) subdomain switching, which removes the implicit trust between pages that happen to belong to the same origin but not necessarily trust each other.

SessionLock [2] negotiates a shared secret between a client and a server and stores this in the client-side context. The secret is used to add integrity checks to outgoing requests. Since the secret value is never transmitted in the clear, SessionLock prevents an attacker with a stolen session identifier from making valid requests. Unfortunately, because the secret is stored in the JavaScript context, it cannot be protected against script-based attacks.

The *HTTP Integrity Header* [25] uses a similar approach as SessionLock but makes the secret negotiation and integrity check part of HTTP protocol, thereby avoiding modifications to the application logic. *SecSess* [15] further improves on HTTP integrity header by achieving compatibility with commonly deployed middleboxes such as Web caches. *GlassTube* [26] also ensures integrity on the data transfer between client and server and can be deployed both within an application or as a modification of the client-side environment, for example as a browser plugin.

Finally, several approaches look into strengthening cookies to prevent session hijacking attacks. *One-Time Cookies* [12] propose to replace the static session identifier with disposable tokens per request, similar to the concept of Kerberos service tickets. Each token can only be used once, but using an initially shared secret, every token can be separately verified and tied to an existing session. *Macaroons* [7] improve upon cookies by placing restrictions on how, where, and when the implicit authority of the bearer token can be used. The technology behind macaroons is based on chains of nested hash-based message authentication codes (HMACs), built from a shared secret and a chain of messages. Macaroons target cloud services, where delegation between principals without a central authentication service is often required, for example to share access to the user's address book on another service.

Other techniques follow a similar approach but base their security measures on the user's password, which in itself is a shared secret between the user and the Web application. *BetterAuth* [33] revisits the entire authentication process, offering secure authentication and a secure subsequent session. *Hardened Stateless Session Cookies* [37] use unique cookie values, calculated using hashing functions based on the user's password, effectively preventing the generation of new requests within an authenticated session.

Alternatively, *origin-bound certificates* (OBC) [19] extend the transport layer security (TLS) protocol to establish a strong authentication channel between browser and server, without falling prey to active network attacks. Within this secure channel, TLS–OBC supports the binding of cookies and third-party authentication tokens, which prevents the stealing of such bearer tokens.

Another line of research targets session hijacking problems from the client side without explicit support from the target application. *SessionShield* [39] is a client-side proxy that mitigates script-based session hijacking attacks by ensuring that all session cookies are marked *HttpOnly* before they reach the browser. Determining which cookies are session cookies at the client side, in an application-agnostic way, is achieved by applying sensible heuristics, including an entropy test. *Serene* [13] implements SessionShield as a browser extension for Firefox and extends it to support parameter-based session management techniques.

7.1.3 State of Practice

Unfortunately, many sites still use unprotected cookies to store session identifiers, leaving users vulnerable to session hijacking attacks. On the bright side, the adoption of the *HttpOnly* and *Secure* attributes is gaining ground, starting to be turned on by default [4]. Our July 2014 survey of the Alexa top 10,000 domains shows that more than half of the domains use the *HttpOnly* attribute (5,465 domains in total) and 1,419 domains use the *Secure* attribute, which is a significant increase compared to a study in 2010 [49].

Another practice that is being deployed by major sites is to operate split session management between HTTP- and HTTPS-accessible parts of the site. For example, a Web shop can run its catalog inspection and shopping cart filling operations over HTTP and use HTTPS for sensitive operations such as logging in, checking out the cart, payments, or account modifications. Technically, they use two different session cookies, one for HTTP usage and another for HTTPS usage, where the latter is declared *HttpOnly* and *Secure*. While this leaves the user vulnerable to a session hijacking attack on HTTP part, it effectively protects HTTPS part, where sensitive operations are conducted.

7.1.4 Best Practices

The best practice for preventing session hijacking attacks is to use strong, random session identifiers [48], and deploy the application over HTTPS (see best practices in Chap. 5), using the *HttpOnly* and *Secure* attribute for all cookies not needed by JavaScript, especially the cookies containing a session identifier.

In addition, Web development frameworks and application servers that offer easy-to-use session management mechanisms should deploy these protections by default, and discourage their users from turning them off.

7.2 Session Fixation

A session fixation attack enables an attacker to force the victim's browser to use an existing session, which is also known by the attacker. The goal of the attacker is to wait for the user to perform state-changing actions, such as authenticating himself to the application, after which the attacker takes control of the session. The effects of a session fixation attack are similar to those of a session hijacking attack.

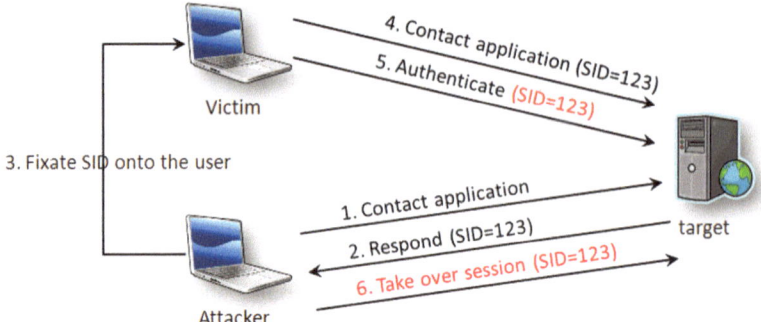

Fig. 7.2 In a *session fixation* attack, an attacker fixates his own session identifier into the browser of the user (step 4), causing the user to authenticate in the attacker's session

Session fixation is categorized as a session management problem, ranked second in the OWASP top ten [51]. Session fixation is technically more difficult than session hijacking, and requires the capability to transfer a session identifier towards the victim's browser. Unfortunately, no exact numbers of the prevalence of session fixation attacks are available. However, the prevalence of the attack vectors that can lead to a session fixation attack is a good indicator, and is covered in subsequent chapters.

7.2.1 Description

The goal of a session fixation attack is to register the results of the victim's state-changing actions in a session controlled by the attacker. The most prominent example of such a state-changing action is the authentication process, which results in the authentication state being stored in the session. The attacker forces the victim's browser to use a specific, attacker-known session, allowing him to retake control of this session at any time. If the attacker takes over the session after user authentication, he can effectively impersonate the user.

For cookie-based session management systems, the attacker first obtains a valid session identifier for the application, either by visiting the target application himself or by crafting a session identifier. In the next step, the attacker has to fixate the session identifier in the victim's browser, which depends on the session management mechanism used by the application. Once the session is fixated and the user visits the application, he will be working within the attacker's session. This means that the authentication state at the server side will be stored within this session as well, allowing the attacker to take over the session later on (illustrated in Fig. 7.2).

The crucial part of a session fixation attack is fixating the session identifier, an action that depends on the presence of a secondary vulnerability such as cross-site scripting, header injection, etc. [13]. For example, in cookie-based session management systems, the attacker can set a cookie using the *document.cookie* property from JavaScript or by using an injection attack to insert *meta* elements that mimic header operations into the page's content or by manipulating network traffic.

Fixating a session identifier in parameter-based session management systems is straightforward. All it takes is tricking the user into visiting a URI which contains the fixated session identifier as a parameter in the query string.

In essence, session fixation attacks are possible because the session identifier acts as a bearer token for a session between a user and an application, combined with the fact that sessions are easily transferable between browsers. Session fixation and session hijacking attacks both exist for the same reasons, but use a different attack vector to obtain an authenticated session.

7.2.2   Mitigation Techniques

Due to multiple attack vectors that can lead to a session fixation attack, plugging them all is difficult. Nonetheless, protecting session cookies with the *Secure* and *HttpOnly* attributes makes the attack more difficult, since it prevents an attacker from easily overwriting an already-existing session cookie. However, these protections can be bypassed, for example, by overflowing the cookie jar with meaningless cookies, causing the browser to purge the oldest ones (i.e., the session cookie) and allowing the attacker to fixate a new session identifier.

An effective mitigation technique for fixation attacks consists of sending the user a renewed session identifier after the user changes privilege levels in the application, such as a login or logout operation, accessing an administrative part of the application, etc. For example, by issuing a new session identifier after user authentication, an application ensures that the authentication information is not associated with the fixated session identifier, preventing the attacker from taking over the authenticated session. Renewing the session identifier is the server's responsibility and is often supported by the Web programming language or Web framework. No explicit client support is required, since the server can just override the already-existing session cookie using a *Set-Cookie* header.

However, integrating the renewal of the session identifier in legacy applications is challenging. Research proposals propose several solutions to this problem, both from the server side and the client side. Depending on the available frameworks at the server side, renewing the session identifier can be integrated in the framework's session management mechanism or be offered as a server-side reverse proxy solution [32]. On the other hand, a client-side protection mechanism against session fixation attacks, called *Serene*, offers protection to a user, without requiring a change or modification at the server side [13]. *Serene* is a browser add-on that detects cookie and parameter-based session identifiers in requests and responses and offers additional protection for these identifiers. *Serene* can prevent fixation attacks initiated by script or metatag injection attacks, as well as by parameter-based session fixation attacks.

A variant of a session fixation attack can be carried out by a *related domain attacker* [8], who controls an application hosted under the same registered domain as the target application (e.g., *example.com*). By setting a cookie that applies to all sibling domains, an attacker can easily fixate a session identifier. *Origin Cookies* [8] protect applications against this kind of attack by allowing cookies to be limited to one domain, preventing manipulation from sibling sites. TLS-OBC [19] is follow-up research that offers even stronger guarantees but requires a TLS-secured channel to be present.

7.2.3 State of Practice

Modern frameworks support the renewal of the session identifier, albeit only after explicit actions from the developer to enable this behavior [47]. In addition, correctly enabling the *HttpOnly* attribute on session cookies can successfully mitigate certain attack vectors.

7.2.4 Best Practices

The best practice for protection against session fixation attacks is to renew the session identifier on every privilege change within the application. This effectively ensures that the new privilege level is never accessible using the original session identifier, thus preventing session fixation attacks. In addition, session cookies should always be issued with the *HttpOnly* attribute set, which prevents overwriting from headers and from JavaScript in most modern browsers.

7.3 Authenticating With Stolen Credentials

Stealing the victim's credentials for a target application allows an attacker to authenticate himself to the application as if he were the user. The attacker can successfully impersonate the user and can also bypass reauthentication checks for sensitive operations such as changing the password. With the prevalent reuse of the same credentials for multiple applications and the use of single-sign on solutions, stealing the credentials for one application often gives the attacker access to other applications as well.

In recent years, credential theft has become a common practice. The compromises of credential databases have become enormous, with the theft of thousands, even millions of users being no exception. Some examples are the 2013 Adobe breach, resulting in the theft of 2.9 million customer details [29], or the compromise of eBay, which asked its 145 million registered users to change their credentials [43]. In addition, attackers can use phishing techniques, where unsuspecting users are tricked into entering their credentials into a fraudulent authentication form. PhishTank, an antiphishing initiative, collects about 20,000 valid phishing Web sites per month [41].

7.3.1 Description

By using valid user credentials, the attacker can impersonate the user towards the target application, resulting in full control over the victim's data and actions. In the modern, interconnected Web, compromise of one account often allows the escalation towards other accounts and the victim's entire online presence [27, 28].

Attackers often employ social engineering techniques to trick victims into willingly surrendering their credentials. The most common example of such an attack is *phishing*, where the attacker capitalizes on a user's inability of distinguishing a legitimate page from one that looks legitimate but is actually fraudulent. By luring the user to the fraudulent page, for example with a carefully crafted "urgent" email message, the user is tricked into entering his credentials, causing them to be sent to the attacker. Phishing attacks can be conducted both on large and small scale, depending on an attacker's objectives. Large-scale attacks are very generic and generally easier to detect. Small-scale attacks, also known as *spear phishing*, target highly specific individuals and companies and are very difficult to detect.

A variation on the traditional phishing attack is *tabnabbing* [42]. In tabnabbing (shown in Fig. 7.3), the user is lured into visiting a malicious site, that however looks innocuous. If a user keeps the attacker's site open and uses another tab of his browser to browse a different Web site, the tabnabbing page takes advantage of the user's lack of focus (accessible through JavaScript as *window.onBlur*) to change its appearance (page title, favicon, and page content) to look identical to the login screen of a popular site. When a user returns back to the open tab, he has no reason to reinspect the URL of the site rendered in it, since he already did that in the past. This type of phishing separates the visit of a site from the actual phishing attack and could, in theory, even trick users who would not fall victim to traditional phishing attacks.

Instead of directly targeting the victim users, attackers can also focus on the target application itself. A common way of obtaining valid credentials for a target application is by compromising the application's database. The database not only contains full details of the registered users but also contains all user credentials. In addition, with the numerous applications requiring authentication credentials, users often reuse the same set of credentials, allowing an attacker to impersonate the victim towards other applications as well.

In essence, attacks based on credential theft are possible because an attacker can easily get hold of a victim's credentials. Traditional username/password-based credentials are easily transferable, often reused, and stored insecurely. In addition, users are not trained security professionals and often fall for well-conducted social engineering attacks.

7.3.2 Mitigation Techniques

A recent evolution towards limiting the impact of credential theft is the use of multifactor authentication. In a multifactor authentication process, the application no

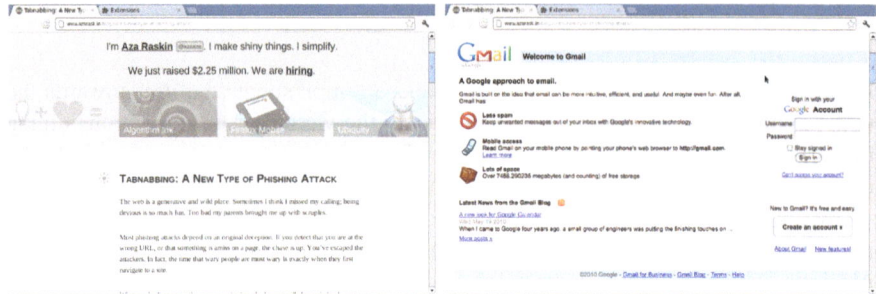

Fig. 7.3 In a *tabnabbing* attack, the attacker switches an innocuous-looking tab (*left*) to a phishing page (*right*), avoiding being caught when the user checks the URI of a newly loaded tab

longer depends on a single piece of knowledge, such as a set of credentials, but requires additional factors, such as a token sent to a user's phone by text message, a token generated by a dedicated device [22], a smart card, biometric information, etc. Multifactor authentication makes the traditional credentials less valuable, since one of the additional authentication factors is an out-of-band device, beyond the control of an attacker. However, introducing additional authentication factors also introduces additional concerns. For example, if the user's smartphone acts as a second factor in the authentication process, a problem arises when the phone is stolen, since it provides both, the browser with potentially stored credentials and the out-of-band device. Similarly, biometrics are often considered a viable alternative to password authentication [5, 6] but they possess different characteristics compared to traditional credentials. For example, fingerprints are left behind everywhere and the readers can easily be fooled [44]. In addition, the amount of biometric information is limited (i.e., ten fingerprints) and revocation is rather difficult.

In addition to multifactor authentication, major sites further improve their authentication procedures with additional security checks when logging in from an untrusted device, similar to anomaly-based prevention of credit card fraud. Microsoft, Facebook, and Google allow you to register trusted computers from where a traditional username/password-based authentication can be used. All other machines require two-factor authentication with a verification code [10, 24].

Currently, several client-side tools are available to store a user's password [36, 45], no longer requiring the user to the remember all accounts and associated passwords. Such tools enable the use of unique, application-specific passwords, limiting the harmful effect of credential theft at the server side. Implementing a safe browser extension for managing passwords is a non-trivial task, as illustrated by research [45] and disclosed vulnerabilities [34].

Attackers have been trying to convince users to voluntarily give up their credentials for the past 19 years [35]. Several studies have been conducted trying to identify why users fall victim to phishing attacks [18, 21] and various solutions have been suggested, such as the use of per-site "page-skinning" [17], security toolbars [52], images [3], trusted password windows [46], use of past-activity knowledge [38], and automatic analysis of the content within a page [53]. Finally, users can also install client-side countermeasures to protect themselves phishing [11] and tabnabbing [14].

7.3.3 State of Practice

In practice, stolen credentials are a valuable asset, as illustrated by the high demand on underground markets [30]. Major Web sites offer strong, multifactor authentication, in combination with trusted devices, which effectively mitigates most of the risk associated with credential theft. Also, major players, such as Google and Facebook offer single-sign on solutions, allowing other sites to benefit from the secure authentication procedures. On the downside, numerous smaller sites still use traditional credentials and cannot prevent the use of stolen credentials.

Unfortunately, combating phishing in an automated way is difficult, which is why the currently deployed antiphishing mechanisms in popular browsers are all blacklist-based [16]. The blacklists themselves are either generated automatically by automated crawlers, searching for phishing pages on the Web [50] or are crowdsourced [41].

7.3.4 Best Practices

The best protection against the use of stolen credentials is multifactor authentication. Instead of building your own multifactor authentication system, you can choose two strategies that involve a third-party authentication provider. The first strategy integrates an additional authentication factor in your own authentication procedure, using APIs offered by third-party providers. The second strategy fully outsources the authentication procedure to an authentication provider, for example, using a protocol like OpenID or OAuth.

Specifically for phishing, a social engineering attack, users should be trained to recognize phishing scams and never act on them. However, as the attacks become more complicated, tool support for detecting and preventing phishing attacks is necessary. The best practice here is the use of safe browsing initiatives, which are based on crowdsourced blacklisting.

References

1. Acar, G., Juarez, M., Nikiforakis, N., Diaz, C., Gürses, S., Piessens, F., Preneel, B.: Fpdetective: dusting the web for fingerprinters. In: Proceedings of the 20th ACM Conference on Computer and Communications Security (CCS), pp. 1129–1140 (2013)
2. Adida, B.: Sessionlock: securing web sessions against eavesdropping. In: Proceedings of the 17th International Conference on World Wide Web (WWW), pp. 517–524 (2008)
3. Agarwal, N., Renfro, S., Bejar, A.: Yahoo!'s sign-in seal and current anti-phishing solutions. Web 2.0 Security and Privacy (W2SP) (2007)

4. Apache Software Foundation: Apache tomcat—migration guide. http://tomcat.apache.org/migration-7.html (2013)
5. Apple: iphone 5s: About touch ID security. http://support.apple.com/kb/HT5949 (2014)
6. Berg, D.: How to use your fingerprint reader. http://blog.laptopmag.com/how-to-use-your-fingerprint-reader (2012)
7. Birgisson, A., Politz, J., Erlingsson, Ú., Taly, A., Vrable, M., Lentczner, M.: Macaroons: cookies with contextual caveats for decentralized authorization in the cloud. In: Proceedings of the 21st Annual Network and Distributed System Security Conference (NDSS) (2014)
8. Bortz, A., Barth, A., Czeskis, A.: Origin cookies: session integrity for web applications. Web 2.0 Security and Privacy (W2SP) (2011)
9. Butler, E.: Firesheep. http://codebutler.com/firesheep (2010)
10. Center, F.H.: Extra security features. https://www.facebook.com/help/413023562082171/ (2014)
11. Chou, N., Ledesma, R., Teraguchi, Y., Mitchell, J.C.: Client-side defense against web-based identity theft. In: Proceedings of the 11th Annual Network and Distributed System Security Conference (NDSS) (2004)
12. Dacosta, I., Chakradeo, S., Ahamad, M., Traynor, P.: One-time cookies: preventing session hijacking attacks with stateless authentication tokens. ACM Trans. Internet Technol. (TOIT) **12**(1), 31 (2012).
13. De Ryck, P., Nikiforakis, N., Desmet, L., Piessens, F., Joosen, W.: Serene: self-reliant client-side protection against session fixation. In: Proceedings of the 12th International IFIP Conference on Distributed Applications and Interoperable Systems (DAIS), pp. 59–72 (2012)
14. De Ryck, P., Nikiforakis, N., Desmet, L., Joosen, W.: Tabshots: client-side detection of tabnabbing attacks. In: Proceedings of the 8th ACM symposium on Information, computer and communications security (ASIACCS), pp. 447–456 (2013)
15. De Ryck, P., Desmet, L., Piessens, F., Joosen, W.: Eradicating bearer tokens for session management. W3C/IAB workshop on strengthening the internet against pervasive monitoring (STRINT) (2014)
16. Developers, G.: Safe browsing API. https://developers.google.com/safe-browsing/ (2014)
17. Dhamija, R., Tygar, J.D.: The battle against phishing: dynamic security skins. In: Proceedings of the 1st Symposium on Usable Privacy and Security (SOUPS), pp. 77–88 (2005)
18. Dhamija, R., Tygar, J.D., Hearst, M.: Why phishing works. In: Proceedings of the ACM CHI conference on Human Factors in computing systems (CHI), pp. 581–590 (2006)
19. Dietz, M., Czeskis, A., Balfanz, D., Wallach, D.S.: Origin-bound certificates: a fresh approach to strong client authentication for the web. In: Proceedings of the 21st USENIX Security Symposium, pp. 16–16 (2012)
20. Eckersley, P.: How unique is your web browser? In: Proceedings of the 10th Privacy Enhancing Technologies Symposium (PETS), pp. 1–18 (2010)
21. Egelman, S., Cranor, L.F., Hong, J.: You've been warned: an empirical study of the effectiveness of web browser phishing warnings. In: Proceedings of the ACM CHI conference on Human Factors in computing systems (CHI), pp. 1065–1074 (2008)
22. EMC: RSA SecurID—Two-Factor Authentication Security Token. http://www.emc.com/security/rsa-sccurid.htm (2013)
23. Geier, E.: Prevent wi-fi eavesdroppers from hijacking your accounts. http://www.ciscopress.com/articles/article.asp?p=1750204 (2011)
24. Google: Trusted computers. https://support.google.com/accounts/answer/2544838?hl=en (2014)
25. Hallam-Baker, P.: Http integrity header. IETF Internet Draft (2012)
26. Hallgren, P.A., Mauritzson, D.T., Sabelfeld, A.: Glasstube: a lightweight approach to Web application integrity. In: Proceedings of the 8th ACM SIGPLAN workshop on Programming Languages and Analysis for Security (PLAS), pp. 71–82 (2013)
27. Hiroshima, N.: How i lost my $50,000 twitter username. https://medium.com/@N/how-i-lost-my-50-000-twitter-username-24eb09e026dd (2014)

28. Honan, M.: How apple and amazon security flaws led to my epic hacking. http://www.wired.com/2012/08/apple-amazon-mat-honan-hacking/ (2012)

29. Infosecurity: Adobe hacked customers' card details and adobe source code stolen. http://www.infosecurity-magazine.com/view/34872/adobe-hacked-customers-card-details-and-adobe-source-code-stolen (2013)

30. Infosecurity: 360 million stolen credentials and 1.25 billion email addresses found on the black market. http://www.infosecurity-magazine.com/view/37135/360-million-stolen-credentials-and-125-billion-email-addresses-found-on-the-black-market/ (2014)

31. Johns, M.: Sessionsafe: implementing xss immune session handling. In: Proceedings of the 11th European Symposium on Research in Computer Security (ESORICS), pp. 444–460 (2006)

32. Johns, M., Braun, B., Schrank, M., Posegga, J.: Reliable protection against session fixation attacks. In: Proceedings of the 26th ACM Symposium on Applied Computing (SAC), pp. 1531–1537 (2011)

33. Johns, M., Lekies, S., Braun, B., Flesch, B.: Betterauth: web authentication revisited. In: Proceedings of the 28th Annual Computer Security Applications Conference (ACSAC), pp. 169–178 (2012)

34. Kelly, S.M.: LastPass passwords exposed for some internet explorer users. http://mashable.com/2013/08/19/lastpass-password-bug/ (2013)

35. Langberg, M.: Aol acts to thwart hackers. http://simson.net/clips/1995/95.SJMN.AOL_Hackers.html (1995)

36. LastPass.com: LastPass. https://lastpass.com (2013)

37. Murdoch, S.J.: Hardened stateless session cookies. Secur. Protoc. **XVI**, 93–101 (2011)

38. Nikiforakis, N., Makridakis, A., Athanasopoulos, E., Markatos, E.P.: Alice, what did you do last time? fighting phishing using past activity tests. In: Proceedings of the 3rd European Conference on Computer Network Defense (EC2ND), pp. 107–117 (2009)

39. Nikiforakis, N., Meert, W., Younan, Y., Johns, M., Joosen, W.: Sessionshield: lightweight protection against session hijacking. In: Proceedings of the 3rd International Symposium on Engineering Secure Software and Systems (ESSoS), pp. 87–100 (2011)

40. Nikiforakis, N., Kapravelos, A., Joosen, W., Kruegel, C., Piessens, F., Vigna, G.: Cookieless monster: exploring the ecosystem of web-based device fingerprinting. In: Proceedings of the 34th IEEE Symposium on Security and Privacy (SP) (2013)

41. OpenDNS: PhishTank. http://www.phishtank.com/ (2014)

42. Raskin, A.: Tabnabbing: a new type of phishing attack. http://www.azarask.in/blog/post/a-new-type-of-phishing-attack/ (2010)

43. Reisinger, D.: eBay hacked, requests all users change passwords. http://www.cnet.com/news/ebay-hacked-requests-all-users-change-passwords/ (2014)

44. Roberts, P.F.: 7 ways to beat fingerprint biometrics. http://www.itworld.com/slideshow/120606/7-ways-beat-fingerprint-biometrics-374041 (2013)

45. Ross, B., Jackson, C., Miyake, N., Boneh, D., Mitchell, J.C.: Stronger password authentication using browser extensions. In: Proceedings of the 14th USENIX Security Symposium (2005)

46. Sandler, D.R., Wallach, D.S.: <input type="password"> must die! Web 2.0 Security and Privacy (W2SP) (2008)

47. Siles, R.: Session management cheat sheet—renew the session id after any privilege level change. https://www.owasp.org/index.php/Session_Management_Cheat_Sheet#Renew_the_Session_ID_After_Any_Privilege_Level_Change (2013)

48. Siles, R.: Session management cheat sheet—session id properties. https://www.owasp.org/index.php/Session_Management_Cheat_Sheet#Session_ID_Properties (2013)

49. Singh, K., Moshchuk, A., Wang, H.J., Lee, W.: On the incoherencies in web browser access control policies. In: Proceedings of the 31st IEEE Symposium on Security and Privacy (SP), pp. 463–478 (2010)

50. Wenyin, L., Huang, G., Xiaoyue, L., Min, Z., Deng, X.: Detection of phishing webpages based on visual similarity. Special Interest Tracks and Posters of the 14th International Conference on World Wide Web (WWW), pp. 1060–1061 (2005)

51. Wichers, D.: Owasp top 10. https://www.owasp.org/index.php/Category:OWASP_Top_Ten_ Project (2013)
52. Wu, M., Miller, R.C., Garfinkel, S.L.: Do security toolbars actually prevent phishing attacks? In: Proceedings of the ACM CHI Conference on Human Factors in Computing Systems (CHI), pp. 601–610 (2006)
53. Zhang, Y., Hong, J.I., Cranor, L.F.: Cantina: a content-based approach to detecting phishing web sites. In: Proceedings of the 16th International Conference on World Wide Web (WWW), pp. 639–648 (2007)

Chapter 8
Attacks on the Client-Side Context

In previous chapters, we have shown how the attacker was able to manipulate the user's actions. By attacking the client-side context of a target application, the attacker actually inserts himself into the target application's context in the user's browser. This gives the attacker the power to monitor the user's interactions with the target application, to read and extract data, and to send seemingly legitimate requests to the server-side application.

In this chapter, we look at three attacks that can lead to the attacker controlling the client-side context. The first attack is cross-site scripting (XSS), a very common, well-known attack, where the attacker injects JavaScript code into the victim application. Next, we cover *scriptless attacks*, where content is also injected into the target application, but the content is not script code. Finally, we investigate the dangers of remote script inclusions, which are very common but also prone to compromise.

8.1 Cross-Site Scripting

With an XSS attack, an attacker is able to execute his own JavaScript code within the application's execution context, gaining him the same privileges as the target application code. This exposes all client-side application data, resources, and APIs to the attacker, including the possibility to manipulate and generate legitimate application requests towards the server-side application code.

XSS is a serious problem in the Web, and is highly ranked in both the OWASP top ten of Web application vulnerabilities [42] and the CWE/SANS most dangerous programming errors [22]. Almost every Web application has had a script injection vulnerability at some point, with even serious players such as Google, Facebook, and Twitter not being exempted [43]. Hence, XSS is often referred to as the *buffer overflow of the Web*.

© Philippe De Ryck, Lieven Desmet, Frank Piessens, Martin Johns 2014
P. De Ryck et al., *Primer on Client-Side Web Security*,
SpringerBriefs in Computer Science, DOI 10.1007/978-3-319-12226-7_8

`http://example.com/search.php?q=%3Cscript%3Ealert(%22XSSed!%22)%3C%2Fscript%3E`

Fig. 8.1 When the vulnerable Web application processes this URI, the source of the response will include `<script>alert(''XSSed!'')</script>`, leading to a reflected XSS attack

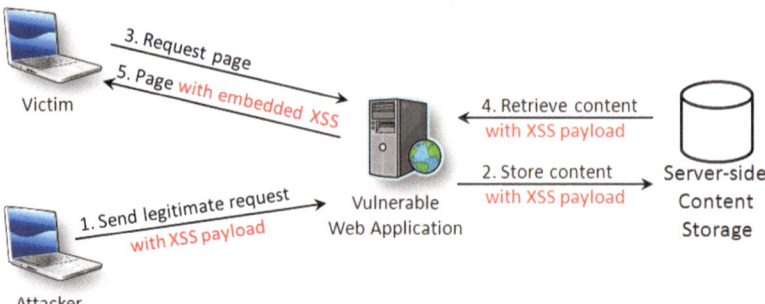

Fig. 8.2 In a stored XSS attack, the attacker injects script code into the application's server-side content storage, which is then unknowingly served to victim users, visiting legitimate pages of the application

8.1.1   Description

The goal of an XSS attack is to execute attacker-controlled code in the client-side application context within the victim's browser. In an XSS attack, the attacker is able to inject JavaScript code into a page of the target application, mixing it with the legitimate page content, causing it to be executed altogether as the page is processed. As the browser sees a single Web page, it is unable to distinguish between legitimate code and malicious code.

An attacker has many attack vectors to inject a payload into the target application. A first way is by manipulating the uniform resource identifier (URI) to inject code into request parameters, which are processed by a client-side script of the target application. Whenever the client-side script constructs code using these parameters, which it does not expect to hold code, the attacker's code will be executed alongside the legitimate application code. This type of XSS attack is known as *document object model (DOM)-based XSS* or *XSS type 0*.

A second class of XSS attacks consists of tricking the server into including the attacker's code in its response. For example, if the attacker makes the victim's browser visit the URI shown in Fig. 8.1, the server will reflect the value of the URI parameter back in the response, where it will be executed as part of the requested page. This type is known as *reflected XSS* or *XSS type 1*.

Finally, an attacker can also store the malicious code in the application's data, for example by hiding in a forum post or blog comment. Whenever the victim requests a page that includes the attacker's content, the malicious code will be embedded in the page as well. This type of XSS is known as *stored XSS* or *XSS type 2*, and is illustrated in Fig. 8.2.

In essence, the problem of an XSS attack is the failure of the target application to recognize the insertion of code, thus allowing the payload to be executed. The combination of the facts that code can be placed anywhere in a document and that browsers attempt to correct syntactically incorrect documents rather than rejecting or ignoring them, helps the easy exploitation of injection vulnerabilities.

8.1.2 Mitigation Techniques

The traditional mitigation technique used against XSS attacks depends on sanitizing input and output, preventing any dangerous input from reaching the final output. These sanitization techniques attempted to simply replace or remove dangerous characters such as < > & " ' or check against a whitelist of allowed characters, but modern sanitization techniques take the context of the output into account.

Modern Web applications generate output for different contexts, with different output formats and injection vectors. Some example contexts are HTML elements, HTML element attributes, cascading style sheets (CSS) code, JavaScript code, etc. Several publicly available libraries provide context-sensitive content encoding, and effectively mitigate XSS attacks. Popular examples of Java applications are the OWASP Java Encoder Project [14], which offers several context-specific sanitization operations, and OWASP's Java XML Templates [13], which offer automatic context-aware encoding. Alternatively, HTML purifier [44] offers automatic sanitization for PHP applications, and even ensures that the output is standards-compliant HTML. Automating context-sensitive sanitization is an active research topic. Script-Gard [33] focuses on the detection of incorrect use of sanitization libraries (e.g., context-mismatched sanitization or inconsistent multiple sanitizations), and is capable of detecting and repairing incorrect placement of sanitizers. Other work focuses on achieving correct, context-sensitive sanitization, using a type-qualifier mechanism to be applied on existing Web templating frameworks [31].

Even with the most advanced mitigation techniques, both newly created and legacy applications remain vulnerable to XSS attacks. Therefore, Mozilla proposed Content Security Policy (CSP) [35], a server-driven, browser-enforced policy to be used as a second line of defense. CSP allows a developer or administrator to strictly define the sources of trusted content, such as scripts, stylesheets, images, etc., preventing the inclusion of malicious scripts from untrusted sources. In addition, CSP prevents the execution of harmful inline content by default. When deploying CSP, a reporting-only mode is available. This mode will report any violations of the policy to the developer, without actually blocking any content. This allows to dry-run a policy before actually deploying it towards users. We give more details about CSP in the next section, when discussing scriptless attacks.

CSP's restrictions on dangerous inline content effectively render-injected script code harmless, since it will not be executed, and the list of trusted sources further limits an attacker when including a remote script file. Currently, CSP is being adopted by major browsers, and is on the standardization track of W3C [36]. One downside of CSP is its impact on an application's code since the application is no longer allowed to use inline code, or dangerous features such as *eval()*. For newly developed applications, this is manageable, but legacy applications require some effort to be made compatible [40]. As a response to this problem, the upcoming 1.1 version of CSP [3] will allow inline scripts if they possess a unique, unguessable nonce. Injected scripts will not be able to provide this nonce, hence will not be executed.

The detection of XSS vulnerabilities in Web applications commonly relies on penetration testing (colloquially referred to as *pentesting*) and static analysis [9, 34]. In addition to these state of practice techniques, the state-of-the-art research focuses on the discovery and detection of potential injection vulnerabilities. Kudzu [32] achieves this using symbolic execution of JavaScript. Gatekeeper [10], on the other hand, allows site administrators to express and enforce security and reliability policies for JavaScript programs and was successfully applied to automatically analyze JavaScript widgets, with very few false positives and no false negatives.

8.1.3   State of Practice

Injection vulnerabilities leading to XSS attacks are prevalent in both new and legacy Web applications. A large-scale analysis of the Alexa top 5,000 sites has discovered 6167 unique XSS vulnerabilities, distributed over 480 domains [19]. XSS attacks are often only the first step in a more complicated attack, involving underlying infrastructure or higher-privilege accounts. The consequences of escalating an XSS attack are aptly demonstrated by exploitation frameworks, such as the Browser Exploitation Framework (BeEF) [2] or Metasploit [29].

Currently, almost every newly developed Web application sanitizes its inputs and outputs, in an attempt to avoid injection vulnerabilities altogether. Most modern development frameworks offer library support for sanitization. Unfortunately, sanitization libraries are not always context-sensitive, and many applications apply sanitization procedures wrongly or inconsistently [33]. In addition, a few context-sensitive sanitization libraries are available, as discussed above as an aspect of mitigation techniques.

Since injection vulnerabilities remain widespread, several attempts have been made to stop them from within the browser, independent of any application-specific mitigation techniques. Examples of in-browser mitigation techniques are XSS filters [4, 30, 37], or the popular security add-on NoScript [21]. The newly introduced CSP [3, 36] is slowly starting to be adopted. Our July 2014 survey of the Alexa top 10,000 sites found 131 sites that already issue a CSP policy in their response headers.

In addition, applications often apply code-based isolation techniques to prevent the damage that can be done by untrusted or injected scripts. Examples of currently available isolation techniques are HTML 5 sandboxes [5] or browser-based sanitization procedures for dynamic script code, such as Internet Explorer's *toStaticHTML()* [6].

8.1.4 Best Practices

The best defense against XSS attacks is to apply proper input and output sanitization. Sanitization has to be context-sensitive, so one should use either sanitization libraries that automatically determine the correct context or use the appropriate sanitization function for the output context at hand.

When filtering input or output, use a whitelist approach, where only the valid patterns are whitelisted, instead of a blacklist approach, which has to be an exhaustive list of prohibited patterns. In addition, avoid writing custom sanitization libraries, which is an error-prone process, especially due to different encoding options, browser quirks, and obscure Web features [46].

If possible, use a tight CSP on your site, fully preventing dangerous inline content and strictly limiting the sources of external content. Even when not directly deploying CSP, it might be useful to adapt your code to support CSP in a subsequent incarnation (e.g., do not use inline scripts, *eval()* etc.).

8.2 Scriptless Injection Attacks

With a scriptless injection attack, an attacker is able to manipulate the client-side execution of a target application, without injecting actual script code. Due to the scriptless characteristic of the attack, it is likely to bypass any XSS mitigation technique. Scriptless attacks may enable an attacker to steal a user's password, extract security tokens from the page, or change the destination of a page's form, but a successful scriptless attack requires a lot of specific circumstances to line up.

8.2.1 Description

Scriptless injection attacks are essentially the same as XSS attacks, with the main difference that the injected payload is not JavaScript code. By not injecting script code, this attack succeeds in bypassing numerous filters and other mitigation techniques.

In a scriptless injection attack, an attacker has a wide variety of content types to choose from. One attack vector is the injection of HTML content [8, 45], allowing an attacker to modify the destination of forms, extract hidden security tokens, etc. For example, by injecting a new *button* element, with a *formaction* attribute to change the destination of a form, an attacker can trick the user into submitting a form towards an attacker-controlled URI. An alternative attack vector is the injection of CSS style code and scalable vector graphics (SVG) image code, allowing an attacker to even extract passwords from an input field [11].

In essence, preventing scripts from being injected does not solve the underlying issue, which is injection vulnerability. By carefully selecting the payload, an attacker is still able to execute sensitive operations, even though he needs to be a bit more creative than with a straightforward XSS attack.

8.2.2 *Mitigation Techniques*

Similar to XSS, mitigating scriptless attacks depend on strict input validation and output sanitization. As these have already been covered, we are not going to repeat them here. Instead, we will elaborate a bit on a promising mitigation technique, CSP [3, 36], as this newly introduced policy is able to prevent most scriptless injection attacks as well.

With CSP, a Web application can set a policy that specifies the characteristics of the page, and where content is loaded from. CSP policies are added to a Web document through an HTTP header or a meta-tag.

One of the goals of CSP is to prevent injected JavaScript from being executed. To achieve this goal, a CSP policy can:

1. Disallow the mixing of HTML mark-up and JavaScript syntax in a single document (i.e., forbidding inline JavaScript, such as event handlers in element attributes).
2. Prevent the runtime transformation of string-data into executable JavaScript via functions such as `eval()`.
3. Provide a list of Web hosts, from which script code can be retrieved.

If used in combination, these three capabilities lead to an effective thwarting of the vast majority of XSS attacks. Forbidding inline scripts renders direct injection of script code into HTML documents impossible. Furthermore, the prevention of interpreting string data as code removes the danger of DOM-based XSS. Finally, only allowing code from whitelisted hosts to run reduces the adversary's capabilities to load custom attack code from external Web locations. Compromising the code on a whitelisted host remains a potential attack vector, even with a tight CSP policy in place.

In addition to putting constraints on the JavaScript that is executed, CSP also limits the CSS code allowed to be processed. CSP defines constraints on the source of stylesheets, and prevents inline stylesheet code from being executed. This already stops several of the attack vectors used by scriptless attacks.

CSP will become even more versatile. In the upcoming 1.1 version [3], a developer will be able to specify where forms can be submitted to, preventing an attacker from overriding the form's action with injected HTML code. Similarly, CSP also enables a developer to define the list of destinations to be used by the XMLHttpRequest object, thus preventing unwanted connections to originate from JavaScript.

For completeness, we also want to mention the other capabilities of CSP. CSP allows the developer to constrain the source of several content types, such as images, media files, embedded objects, fonts, etc. The upcoming 1.1 version will also provide a way to sandbox the page, enable a browser-provided XSS filter to mitigate reflected XSS attacks, and eases the developer's life by offering a nonce-based system to selectively enable inline scripts.

In our social network example application (Chap. 1), we include gadgets from third-party providers. When these are loaded from anywhere on the Web, CSP would need to be very open, as scripts can be loaded from anywhere. However, when we create a gadget store where third-party providers can upload their gadgets, CSP in Listing 8.1 would only allow scripts from our own application, our subdomains, which are the commercial spaces, and the gadgets store. We limit all other content-types to our own domain, except for images, which can be loaded from anywhere. Whenever a CSP violation is detected, it is reported to our violation handler, located at the given URI.

```
Content-Security-Policy:
  default-src 'self';
  img-src *;
  script-src 'self' *.oursocialnetwork.com *.oursocialgadgets.com;
  report-uri http://csp.oursocialnetwork.com/report/;
```

Listing 8.1 This Content Security Policy header locks down the social networking example application introduced in Chap. 1. Violations are reported to the handler located at the given reporting URI

8.2.3 Best Practices

Similar to XSS, the best practices to defend against scriptless attacks are properly applied input and output sanitization. As a second layer of defense, a strict CSP can severely limit the consequences of an attacker exploiting a remaining injection vulnerability.

8.3 Compromised Script Inclusions

Compromising an included JavaScript file allows an attacker to execute attacker-controlled code within the execution context of the application, giving him the same level of privilege as the application code itself. The capabilities gained by such an attack are essentially the same as with an XSS attack, but the attack vector and required capabilities vastly differ.

The inclusion of third-party JavaScript code is a common practice on the Web. Examples are the use of popular libraries and the inclusion of advertisement code. By including remote scripts, the target application trusts these providers to be secure, and offer non-malicious code. Unfortunately, in practice, this is not the case. For example, in 2014, the Reuters Web site got compromised because of an insecure advertisement provider [16]. Advertisement networks are known to let malicious code slip through every now and then [17].

8.3.1 *Description*

The goal of compromising the inclusion of a piece of JavaScript is to gain control over the client-side context. Since included JavaScript code tends to run within the security context of the including page, it suffices for the attacker to compromise any of the included scripts. One example that achieves this goal is to compromise a popular JavaScript library, hosted on a third-party server and included by many Web applications. Note that while a compromised script inclusion yields a similar result as an XSS attack, the required attacker capabilities are vastly different. While an XSS attack can be carried out by a forum poster, compromising a script inclusion requires a network-level attacker, or a server-side attacker who has gained control over a third-party script provider.

Since included scripts can come from a variety of sources, such as a third-party provider, a remote server, or the local storage facility within the browser, an attacker can compromise such a script inclusion in different ways. Target applications that depend on third-party script providers, such as libraries or advertisements, are vulnerable to a compromise of the third-party code. An attacker can compromise a remote machine or attempt to provide malicious content to be distributed, such as an advertisement [16]. Alternatively, an attacker that can monitor and manipulate network traffic, for example, on a publicly accessible wifi network, can manipulate requests and responses that include remote scripts. Manipulation of script files on the network is particularly dangerous for mixed content applications, which are deployed over HTTPS but include scripts from HTTP resources. These scripts are vulnerable to network attackers, violating the security guarantees offered by the secure deployment. Finally, target applications that load script code from a storage facility within the browser, such as the Web Storage API [12], are vulnerable to an attacker that compromises the stored data.

In essence, the inclusion of compromised JavaScript is a problem of code integrity, where the browser lacks the ability to verify that the included code corresponds to the expected code and delivers precisely the required functionality without malicious additions.

8.3.2 *Mitigation Techniques*

The key to mitigating the compromise of script files stored on the server lies in protecting the server and its application against potential adversaries. For applications that require a high degree of control over these script files, it can be useful to copy the third-party script files to their own environment, where they can be optimally protected. Naturally, the copy is only a snapshot of the third-party code and should be kept up-to-date with new versions as they are released.

Mitigating the manipulation of script files on the network can be achieved by deploying the application over HTTPS, and only including remote files received over secure connections. *Mixed content* Web pages, where a secure page includes content over insecure HTTP connections, should be avoided at all times, since they would allow total compromise of the secure application context.

Controlled integration of potentially untrusted JavaScript has been extensively researched, resulting in several proposals. Fine-grained control over the behavior of the included script can be achieved using an inline reference monitor through security wrappers in JavaScript [20, 28], using sandboxing techniques that enforce a policy through a traditional reference monitor [1, 15, 24, 26, 38] and the use of a security-enhanced browser [23, 39].

8.3.3 State of Practice

Research focusing on the exploration of current practices offers valuable information on the State of Practice. A study of the JavaScript inclusion behavior of the top 10,000 Alexa sites [27] reveals that 88.45 % of these sites include at least one remote JavaScript library, and some sites trust as many as 295 remote hosts. The study also attempts to characterize the security of both, including Web applications and third-party script providers, showing that about 12 % of Web applications with a high security classification include content from at least one provider with a low security classification.

Another study of the client-side caching of script code on the top 500,000 Alexa sites [18] shows that 386 Web applications use local storage facilities to cache JavaScript, HTML, and CSS code, as well as 68 entries of remote URIs, used to fetch resources during execution. In addition, a mitigation technique for attacks on locally cached code is proposed, based on cryptographic checksums of the stored code.

Unfortunately, many sites use *mixed content*, violating the security guarantees of HTTPS pages. A recent study shows that 26 % of the TLS-protected Alexa Top 100,000 Web sites included JavaScript over HTTP [7]. A positive evolution is that browsers are all deciding to effectively block the loading of insecure content on secure pages, which has in turn led to standardization activities on the blocking of mixed content [41]. Blocking mixed content has been introduced in Internet Explorer 9, Firefox 23, and Chrome 14, but unfortunately, mobile browsers, which make up 16.68 % of the browser market share, mostly allow mixed content to be loaded [7].

8.3.4 Best Practices

Best practices to prevent the compromise of included scripts focus on limiting the number of trusted third-party hosts, as well as securing the remote host containing the scripts, potentially copying third-party scripts to a controlled server. In addition, deploying your application over HTTPS and ensuring that scripts are only included over HTTPS connections can thwart a network adversary.

When storing any code fragment at the client side, the application needs to ensure its integrity before loading it into its execution environment. This can be achieved by using checksums, which are stored and computed independently from the potentially untrusted code.

In the near future, we expect sandboxing technologies, such as Google Caja [26], Secure ECMAScript [25], JS and [1], etc., to improve in performance and become more developer-friendly, making them a viable candidate for isolating potentially untrusted scripts in a fine-grained, controlled manner.

References

1. Agten, P., Van Acker, S., Brondsema, Y., Phung, P.H., Desmet, L., Piessens, F.: JSand: complete client-side sandboxing of third-party JavaScript without browser modifications. In: Proceedings of the 28th Annual Computer Security Applications Conference (ACSAC), pp. 1–10 (2012)
2. Alcorn, W.: Browser exploitation framework (BeEF). http://beefproject.com (2013)
3. Barth, A., Veditz, D., West, M.: Content security policy level 2. W3C Working Draft (2014)
4. Bates, D., Barth, A., Jackson, C.: Regular expressions considered harmful in client-side xss filters. In: Proceedings of the 19th International Conference on World wide W (WWW), pp. 91–100 (2010)
5. Berjon, R., Faulkner, S., Leithead, T., Navara, E.D., O'Connor, E., Pfeiffer, S., Hickson, I.: HTML 5.1 specification — the sandbox attribute. W3C Working Draft (2014)
6. Center, I.E.D.: Making HTML safer: details for toStaticHTML (Windows Store apps using JavaScript and HTML). http://msdn.microsoft.com/en-us/library/ie/hh465388.aspx (2012)
7. Chen, P., Nikiforakis, N., Desmet, L., Huygens, C.: A dangerous mix: large-scale analysis of mixed-content websites. In: Proceedings of the 16th Information Security Conference (ISC) (2013)
8. De Ryck, P., Desmet, L., Philippaerts, P., Piessens, F.: A security analysis of next generation web standards. Tech. rep., European Network and Information Security Agency (ENISA) (2011)
9. Fergal Glynn, V.: Static code analysis. http://www.veracode.com/security/static-code-analysis (2013)
10. Guarnieri, S., Livshits, V.B.: GATEKEEPER: mostly static enforcement of security and reliability policies for JavaScript code. In: Proceedings of the 18th USENIX Security Symposium, pp. 151–168 (2009)
11. Heiderich, M., Niemietz, M., Schuster, F., Holz, T., Schwenk, J.: Scriptless attacks: stealing the pie without touching the sill. In: Proceedings of the 19th ACM Conference on Computer and Communications Security (CSS), pp. 760–771 (2012)
12. Hickson, I.: Web storage. W3C Recommendation (2013)
13. Ichnowski, J., Manico, J.: Owasp's java xml templates. http://code.google.com/p/owasp-jxt/ (2013)
14. Ichnowski, J., Manico, J., Long, J.: Owasp java encoder project. https://www.owasp.org/index.php/OWASP_Java_Encoder_Project (2013)
15. Ingram, L., Walfish, M.: Treehouse: Javascript sandboxes to help web developers help themselves. In: Proceedings of the USENIX Annual Technical Conference (ATC) (2012)
16. Jacobs, F.: How reuters got compromised by the syrian electronic army. https://medium.com/@FredericJacobs/the-reuters-compromise-by-the-syrian-electronic-army-6bf570e1a85b (2014)

17. Kirk, J.: Yahoo's malware-pushing ads linked to larger malware scheme. http://www.pcworld.com/article/2086700/yahoo-malvertising-attack-linked-to-larger-malware-scheme.html (2014)

18. Lekies, S., Johns, M.: Lightweight integrity protection for web storage-driven content caching. Web 2.0 Security and Privacy (W2SP) (2012)

19. Lekies, S., Stock, B., Johns, M.: 25 million flows later: large-scale detection of dom-based xss. In: Proceedings of the 20th ACM Conference on Computer and Communications Security (CCS), pp. 1193–1204 (2013)

20. Magazinius, J., Phung, P.H., Sands, D.: Safe wrappers and sane policies for self protecting javascript. In: Proceedings of the 15th Nordic Conference on Secure IT Systems (NordSec), pp. 239–255 (2010)

21. Maone, G.: NoScript - JavaScript/Java/Flash blocker for a safer Firefox experience! http://noscript.net/ (2013)

22. Martin, B., Brown, M., Paller, A., Kirby, D.: Cwe/sans top 25 most dangerous programming errors. http://cwe.mitre.org/top25/ (2011)

23. Meyerovich, L., Livshits, B.: ConScript: specifying and enforcing fine-grained security policies for Javascript in the browser. In: Proceedings of the 31st IEEE Symposium on Security and Privacy (SP), pp. 481–496 (2010)

24. Mickens, J.: Pivot: fast, synchronous mashup isolation using generator chains. In: Proceedings of the 35th IEEE Symposium on Security and Privacy (SP), pp. 261–275 (2014)

25. Miller, M.S.: Secure EcmaScript 5. http://code.google.com/p/es-lab/wiki/SecureEcmaScript (2011)

26. Miller, M.S., Samuel, M., Laurie, B., Awad, I., Stay, M.: Caja: safe active content in sanitized javascript. http://google-caja.googlecode.com/files/caja-spec-2008-01-15.pdf (2008)

27. Nikiforakis, N., Invernizzi, L., Kapravelos, A., Van Acker, S., Joosen, W., Kruegel, C., Piessens, F., Vigna, G.: You are what you include: large-scale evaluation of remote javascript inclusions. In: Proceedings of the 19th ACM Conference on Computer and Communications security, pp. 736–747 (2012)

28. Phung, P.H., Sands, D., Chudnov, A.: Lightweight self-protecting Javascript. In: Proceedings of the 4th ACM Symposium on Information, Computer and Communications Security (ASIACCS), pp. 47–60 (2009)

29. Rapid7: Metasploit. http://www.metasploit.com/ (2013)

30. Ross, D.: IE 8 XSS Filter Architecture / Implementation. http://blogs.technet.com/b/srd/archive/2008/08/19/ie-8-xss-filter-architecture-implementation.aspx (2008)

31. Samuel, M., Saxena, P., Song, D.: Context-sensitive auto-sanitization in web templating languages using type qualifiers. In: Proceedings of the 18th ACM Conference on Computer and Communications Security (CCS), pp. 587–600 (2011)

32. Saxena, P., Akhawe, D., Hanna, S., Mao, F., McCamant, S., Song, D.: A symbolic execution framework for JavaScript. In: Proceedings of the 31st IEEE Symposium on Security and Privacy (SP), pp. 513–528 (2010)

33. Saxena, P., Molnar, D., Livshits, B.: SCRIPTGARD: automatic context-sensitive sanitization for large-scale legacy Web applications. In: Proceedings of the 18th ACM Conference on Computer and Communications Security (CCS), pp. 601–614 (2011)

34. Security, H.E.: HP fortify static code analyzer (SCA). http://www.hpenterprisesecurity.com/products/hp-fortify-software-security-center/hp-fortify-static-code-analyzer (2013)

35. Stamm, S., Sterne, B., Markham, G.: Reining in the web with content security policy. In: Proceedings of the 19th International Conference on World wide web (WWW), pp. 921–930 (2010)

36. Sterne, B., Barth, A.: Content security policy 1.0. W3C Candidate Recommendation (2012)

37. Stock, B., Lekies, S., Mueller, T., Spiegel, P., Johns, M.: Precise client-side protection against dom-based cross-site scripting. In: Proceedings of the 23rd USENIX Security Symposium, pp. 655–670 (2014)

38. Ter Louw, M., Ganesh, K.T., Venkatakrishnan, V.: AdJail: practical enforcement of confidentiality and integrity policies on Web advertisements. In: Proceedings of the 19th USENIX Security Symposium, pp. 371–388 (2010)
39. Van Acker, S., De Ryck, P., Desmet, L., Piessens, F., Joosen, W.: WebJail: least-privilege integration of third-party components in web mashups. In: Proceedings of the 27th Annual Computer Security Applications Conference (ACSAC), pp. 307–316 (2011)
40. Weinberger, J., Barth, A., Song, D.: Towards client-side html security policies. In: Proceedings of the 6th USENIX Workshop on Hot Topics on Security (HotSec) (2011)
41. West, M.: Mixed content. W3C Working Draft (2014)
42. Wichers, D.: Owasp top 10. https://www.owasp.org/index.php/Category:OWASP_Top_Ten_Project (2013)
43. XSSed: XSS Archive. http://www.xssed.com/archive/ (2014)
44. Yang, E.Z.: HTML Purifier. http://htmlpurifier.org/ (2013)
45. Zalewski, M.: Postcards from the post-xss world. http://lcamtuf.coredump.cx/postxss/ (2011)
46. Zalewski, M.: The Tangled Web: A Guide to Securing Modern Web Applications. San Francisco, No Starch Press (2012)

Chapter 9
Attacks on the Client Device

Previous chapters have covered attacks that have come closer and closer to the victim and increased in impact. In this final attack chapter, we cover attack vectors that lead to a compromise of the user's device. Such a compromise has a high impact, as many of the previously covered countermeasures depend on a trusted environment at the client side, which can no longer be guaranteed if the browser or device is compromised.

We cover two important attack vectors against the user's device. The first one uses drive-by download techniques, where the victim is served malware. The malware in turn exploits a vulnerability in the browser or a plugin such as the Flash player, Java runtime environment, or anything else the user has installed. The second technique uses malicious browser extensions, which commonly have a high degree of control over the client device.

9.1 Drive-By Downloads

Rendering a simple Web page involves a lot of client-side components, such as the browser, the rendering engine, plugins, extensions, all of which can contain vulnerabilities. An attacker looking to compromise a client system can carefully craft specific Web content to exploit such a memory corruption vulnerability, for example, by using a malicious PDF to exploit a buffer overflow in the PDF reader plugin. By putting the exploit software online, and tricking the user into visiting it, the attacker can gain ground on the client machine, allowing him to install malware and further compromise the machine.

Drive-by downloads are a common attack vector on the Web, driven by an entire underground economy of organized crime. Every compromised computer is worth money, or can be used for other criminal activities, such as joining botnets, concealing traffic, etc. The same approach is followed by the National Security Agency (NSA), in their FOXACID program [26], where they aim at compromising client machines for intelligence gathering. Technically, the user is redirected to a FOXACID server, where an algorithm determines the technical skills of the user. Based on their

© Philippe De Ryck, Lieven Desmet, Frank Piessens, Martin Johns 2014   95
P. De Ryck et al., *Primer on Client-Side Web Security,*
SpringerBriefs in Computer Science, DOI 10.1007/978-3-319-12226-7_9

score, they get served a certain kind of malware. Technically savvy users get dumb malware, while the less technically capable users get served the advanced malware. The reasoning behind this decision is that the latter will be less likely to detect the malware, which is an asset that is costly to develop.

9.1.1 Description

In a drive-by download attack, also known as a drive-by exploit attack, a user's computer becomes infected with malware, delivered through the Web platform. By exploiting a client-side memory corruption vulnerability, for example, a buffer overflow vulnerability in the browser, a rendering engine or a browser plugin, an attacker can install malware on the user's machine, giving him full control over the client machine and all the user's actions. Drive-by download attacks are generally executed in such a way that they are completely undetectable by the user. Alternatively, an attacker can attempt to trick the user into explicitly installing malicious software, for example, by masquerading the malware as an antivirus package.

A drive-by download attack happens in several steps. First, the attacker puts the JavaScript code that will trigger the drive-by download on a Web server. He can use his own server, or he can compromise another server through another vulnerability, such as cross-site scripting (XSS), or by using a server-side attack vector, such as structured query language (SQL) injection, command injection, malicious upload, etc. When this JavaScript code is loaded in the victim's browser, it will contact a redirection service that will guide the user towards an appropriate exploit server, depending on the detected operating system, browser version, and available plugins. If the exploit server succeeds in delivering a matching exploit for the targeted vulnerability, the actual malware will be downloaded from a malware server, and installed on the user's machine. Once installed, the malware is generally controlled using a *command and control* system.

Technically, exploiting a vulnerability at the client side can be as simple as serving a specially crafted image, aiming to abuse a vulnerability in the rendering engine [6], or exploiting a vulnerability in a plugin [30]. Using a memory corruption vulnerability, the attacker can jump to the payload in memory, triggering the malicious code to be executed. A common tactic for exploiting a vulnerability is a *heap spraying attack* [14], where the memory is filled with the malicious payload, smoothing out memory alignment issues when jumping to the payload.

In essence, drive-by download attacks are a distributed variant of traditional native code attacks, such as buffer overflow attacks [12]. The exploited code is responsible for processing Web content, which comes from various sources, making it easy to insert malicious content somewhere along the way, thereby exploiting the vulnerable client-side code.

9.1.2 Mitigation Techniques

One technical mitigation technique consists of isolating the plugin execution environment in a process-level sandbox, making it significantly harder to gain full system-level access with a successful exploit against a plugin. This mitigation technique has currently been deployed by all major browsers to sandbox the Flash player, which has a rocky security history on the Web.

A non-technical mitigation technique is to frequently update the installed plugins, in order to benefit from security updates. A significant improvement [11] in this area is the automatic update processes, either deployed by the plugins themselves, or by integrating them into the browser. An example of the former is the Java plugin, which installs its own auto-update mechanism, and an example of the latter is Google Chrome in combination with the Flash plugin, which are bundled and automatically updated when necessary. Automatic updates help mitigate newly developed exploits, often reverse-engineered from the latest security update [27].

The detection and analysis of Web-based malware is an expanding research field, covering the detection and prevention of drive-by download attacks, heap spraying attacks, or the underlying economic models of the malware industry. One line of research focuses on static detection of malware [9, 19], while another relies on feature extraction and classification [8, 25]. Alternative ideas focus on supportive tasks for detection, such as de-cloaking malware [18] or automated collection and replay of malware scenarios [7].

Next to traditional mitigation techniques against native code attacks [32], such as address space layout randomization (ASLR) or data execution prevention (DEP), researchers also investigate these traditional attacks and defenses in the context of the Web, for example, by introducing a new way to execute a heap spraying attack with HTML5 [22]. Runtime monitoring infrastructures are able to detect heap spraying attacks [24], and a modification of the JavaScript engine can completely prevent heap spraying attacks [14].

In addition, security researchers also investigate the underlying economic models of the malware industry. The underground malware economy has evolved quickly, with *pay-per-install* services being offered as a commodity. An extensive study [4] investigates the different families of malware, repacking strategies to avoid detection, and the targeting of specific countries. Another study examining the underground economy of fake antivirus software [29] reveals that three large-scale businesses earned a combined revenue of $ 130 million. Fake antivirus businesses even go a step further, and actively monitor credit card chargebacks of their duped customers. When the number of chargebacks increases, the businesses will grant more refunds without triggering a complaint with the credit card companies, to avoid anomaly detection and remain undetected.

9.1.3   State of Practice

Drive-by download attacks are still on the rise, and are considered an important threat to user's security on the Web. An emerging trend is the shift to single uniform resource identifiers (URIs) that distribute the malicious software, instead of using an entire underlying botnet infrastructure [13]. This shift in distribution mechanism makes lawful takedowns more difficult, as URIs are not that easily blocked, and quickly changed after takedown.

9.1.4   Best Practices

The best practice for Web users is to reduce the number of plugins installed to its absolute minimum and keep all client software up-to-date, including operating system, drivers, browsers, plugins, etc. In addition, the use of browser's *click-to-play* features can reduce the attack surface significantly. In corporate environments, the software on client machines should be controlled, and be kept up-to-date as much as possible.

Web developers should ensure that their applications are well-protected, especially against injection attacks, preventing the leverage of their Web site as a malware distribution platform. Third-party libraries and their providers should be selected carefully, as a compromise of a library provider can also lead to a compromise of all depending Web applications.

9.2   Malicious Browser Extensions

Browser extensions provide additional code that runs within the browser, and has significantly more privileges than traditional Web code. Attackers who are able to compromise legitimate extensions, or trick users into installing malicious extensions, potentially gain the power to inspect and manipulate all the Web applications running within the compromised browsers, and might even be able to compromise the host system of the victim.

Browser extensions have become very common, and almost every Firefox and Chrome user uses them. The official Firefox and Chrome repositories contain thousands of extensions, of which some may be malicious [23], or may turn malicious afterwards. One real-life example of how browser extensions can become malicious was uncovered in 2014 [1]. Adware vendors bought several Chrome extensions for "a four figure number," giving them full control over the extension. They modified the code to publish ads all over the place and pushed an update through the Chrome Web Store, reaching about 30,000 users.

9.2.1   Description

The goal of controlling a browser extension is to have attacker-controlled, privileged code running within the browser. This may give an attacker access to the browser's internal state, and can be an enabling factor allowing escalation of the attack towards full compromise of the client machine. Since browser extensions run on a higher privilege level, outside of the traditional browser security policies, and are able to inspect and manipulate multiple sites, they are an attractive, high-powered target for attackers.

Compromising legitimate browser extensions becomes possible when the extension treats untrusted content carelessly. Since extensions are commonly written in JavaScript, an attacker can perform a script injection attack by manipulating the input, for example, when the extension inspects a page loaded in the browser. Handling such input carelessly results in a script injection attack vector, allowing the attacker to execute arbitrary code within the extension's context (similar to XSS attacks in Chap. 8).

Tricking the user into installing a malicious extension can be done in various ways. The simplest way is to simply provide the extension on a Web site, hoping to trick the user into installing it manually. Another approach is to offer the extension through the official download channels, such as the browser vendor's extension store. Finally, a powerful attacker can spoof the entire extension store, allowing him to have a malicious extension masquerade as a legitimate, popular extension.

In essence, extensions can be compromised when the privileged extension code fails to adequately sanitize untrusted content, allowing an attacker to inject malicious code into the privileged runtime context. The core problem of malicious extensions is social engineering, where users can be tricked into installing an extension from potentially untrusted sources.

9.2.2   Mitigation Techniques

The techniques for mitigating compromised extensions or preventing the installation of malicious extensions are rather limited, and are part of the browser's architecture and the extension store's platform. In general, the consequences of the compromise of a legitimate extension can be addressed by a thorough extension architecture, where code is strictly separated and application programming interface (APIs) are restricted by permissions. Preventing the installation of malicious extensions is not trivial, and can be addressed by preventing installations through unofficial channels, and performing code reviews on the extensions published through the official channel. Each of these techniques is covered in more detail in the *State of Practice*.

Research of the early extension systems and the gaps between required and granted permissions has led to the proposal of an extension system based on the principles of least-privilege, isolated worlds, and permission systems [3], a model that has been adopted as the Google Chrome extension system. Follow-up research [5] investigates the actual Google Chrome extension security architecture in detail, concluding that even with these restrictions in place, many extensions can be compromised by an attacker. As a result, additional defenses have been proposed and deployed, such as the default enforcement of Content Security Policy [28] on Chrome extensions.

Another line of research uses formal systems to verify security properties, often finding and fixing security vulnerabilities in the process. Most research focuses on existing extension systems and extensions, using, for example, information flow analysis to determine whether extensions suffer from privilege escalation [2, 10], or employing type systems to check whether extensions violate the properties of private browsing mode [20]. Other research concludes that the current systems grant too many privileges to an extension, and therefore propose a new extension security model, underpinned by a verification methodology to check an extension's safety [15]. The feasibility of this new extension model has been demonstrated by implementing extensions for popular browsers, including Firefox, Chrome, and Internet Explorer. Finally, in recent research results, an automated way of eliciting malicious behavior in browser extensions is proposed in [17]. The technique uses honeypages and fuzzing to discover malicious behavior and finds several classes of malicious extensions, of which some have over 5.5 million installations.

9.2.3 State of Practice

The state of practice in protecting an extension against compromise, or protecting the users from installing malicious extensions, is defined by the currently available browsers. We cover two modern browsers, Mozilla Firefox and Google Chrome, both of which have extensive support for extensions, and offer a large number of extensions through an official channel.

Firefox's architecture supports privileged extensions, which have access to a large set of browser-provided APIs, offering numerous services, as well as access to the browser's internals and operating system resources, such as reading/writing files, spawning new processes, etc. In Firefox, extensions can define core components, which can be exposed through an API, as well as scripts that interact directly with Web content. In the recently introduced JetPack model [21], extensions can also choose to adopt a more modular model, offering some isolation and restrictions. Concretely, extensions in Firefox are subject to very few limitations, and can easily share their functionality among core components and scripts interacting with Web content.

Mozilla's extension platform, called *Mozilla Add-Ons*, is the official channel to offer extensions to users. Published extensions are guaranteed to have undergone a review by an editor, who is tasked with checking the functionality and behavior of the extension. Firefox also supports the installation of *unofficial* extensions from arbitrary sites but not without the explicit approval of the user.

Chrome's architecture is based on the principles of least privilege, isolated worlds, and permissions. Extensions have a core component, which runs separately from content scripts, which interact with actual Web content. Communication between both contexts is available through the Web Messaging API [16]. In addition, all extensions have a distinct namespace and are isolated from each other. The browser

APIs offered by Chrome are more limited than the Firefox APIs, especially for reaching out of the browser sandbox, into the OS. Furthermore, Chrome extensions have to explicitly request a set of permissions for a determined set of Web sites (wildcards are allowed) upon installation. Without the necessary permissions, several APIs become inaccessible, preventing an extension from escalating its power within the browser.

Chrome's extension platform, called the *Chrome Web Store*, is Chrome's official channel for distributing extensions. Chrome does not perform any reviews, making the Web Store a reputation-based system, where users are expected to file abuse reports in case of a misbehaving extension. Chrome does not support *unofficial* extensions, except from local folders in developer mode. Chrome also disables extensions by default in private browsing mode, called *incognito mode*, since they might be a risk for a user's privacy. They can, however, be explicitly enabled in private browsing mode, if desired.

We conclude with discussing Greasemonkey, essentially an extensible extension. Greasemonkey is a Firefox extension that allows users to run custom scripts on any Web page, allowing them to remove unwanted features from Web applications, or add additional, desired features. Greasemonkey has an associated community-driven script market, hosting more than 140,000 scripts, which was analyzed in a recent study [31]. A malware analysis of 592 scripts labeled as harmful shows that 126 do in fact attempt to steal private data. Further security analysis of 86,358 scripts uncovers 1,736 scripts with document object model (DOM)-based XSS vulnerabilities. In 944 cases, these vulnerabilities could be used by an attacker to trigger an XSS vulnerability on any site, simply by sending the victim a crafted URI.

9.2.4 Best Practices

A best practice for any Web user is to limit the number of extensions to the minimum, and uninstall, or disable those that are not or infrequently needed. In addition, when using a form of private browsing mode, it can be useful to disable extensions, since they can potentially compromise the private nature of the browsing mode [20]. In a corporate environment, it makes sense to prevent the installation of extensions altogether.

References

1. Amadeo, R.: Adware vendors buy Chrome extensions to send ad- and malware-filled updates. http://arstechnica.com/security/2014/01/malware-vendors-buy-chrome-extensions-to-send-adware-filled-updates/ (2014)
2. Bandhakavi, S., King, S.T., Madhusudan, P., Winslett, M.: Vex: vetting browser extensions for security vulnerabilities. In: Proceedings of the 19th USENIX Security Symposium, pp. 339–354 (2010)
3. Barth, A., Felt, A.P., Saxena, P., Boodman, A.: Protecting browsers from extension vulnerabilities. In: Proceedings of the 17th Annual Network and Distributed System Security Conference (NDSS) (2010)

4. Caballero, J., Grier, C., Kreibich, C., Paxson, V.: Measuring pay-per-install: the commoditization of malware distribution. In: USENIX Security Symposium (2011)
5. Carlini, N., Felt, A.P., Wagner, D.: An evaluation of the Google Chrome extension security architecture. In: Proceedings of the 21st USENIX Security Symposium (2012)
6. CERT: Microsoft Internet Explorer buffer overflow in PNG image rendering component. Vulnerability Note VU#189754 (2005)
7. Chen, K.Z., Gu, G., Zhuge, J., Nazario, J., Han, X.: Webpatrol: automated collection and replay of web-based malware scenarios. In: Proceedings of the 6th ACM Symposium on Information, Computer and Communications Security (ASIACCS), pp. 186–195 (2011)
8. Cova, M., Kruegel, C., Vigna, G.: Detection and analysis of drive-by-download attacks and malicious javascript code. In: Proceedings of the 19th International Conference on World Wide Web (WWW), pp. 281–290 (2010)
9. Curtsinger, C., Livshits, B., Zorn, B.G., Seifert, C.: Zozzle: fast and precise in-browser javascript malware detection. In: Proceedings of the 20th USENIX Security Symposium, pp. 33–48 (2011)
10. Dhawan, M., Ganapathy, V.: Analyzing information flow in JavaScript-based browser extensions. In: Proceedings of the 25th Annual Computer Security Applications Conference (ACSAC), pp. 382–391 (2009)
11. Duebendorfer, T., Frei, S.: Why silent updates boost security. Tech. rep., TIK, ETH Zurich (2009)
12. Erlingsson, Ú., Younan, Y., Piessens, F.: Low-level software security by example. In: Handbook of Information and Communication Security, pp. 633–658 (2010)
13. European Union Agency for Network and Information Security (ENISA): ENISA threat landscape, mid-year 2013. https://www.enisa.europa.eu/activities/risk-management/evolving-threat-environment/enisa-threat-landscape-mid-year-2013/ (2013)
14. Gadaleta, F., Younan, Y., Joosen, W.: Bubble: A JavaScript engine level countermeasure against heap-spraying attacks. In: Proceedings of the 2nd International Symposium on Engineering Secure Software and Systems (ESSoS), pp. 1–17 (2010)
15. Guha, A., Fredrikson, M., Livshits, B., Swamy, N.: Verified security for browser extensions. In: Proceedings of the 32nd IEEE Symposium on Security and Privacy (SP), pp. 115–130 (2011)
16. Hickson, I.: HTML5 web messaging. W3C Candidate Recommendation (2012)
17. Kapravelos, A., Grier, C., Chachra, N., Kruegel, C., Vigna, G., Paxson, V.: Hulk: eliciting malicious behavior in browser extensions. In: Proceedings of the 23rd USENIX Security Symposium, pp. 641–654 (2014)
18. Kolbitsch, C., Livshits, B., Zorn, B., Seifert, C.: Rozzle: De-cloaking internet malware. In: Proceedings of the 33rd IEEE Symposium on Security and Privacy (SP), pp. 443–457 (2012)
19. Laskov, P., Šrndić, N.: Static detection of malicious javascript-bearing pdf documents. In: Proceedings of the 27th Annual Computer Security Applications Conference (ACSAC), pp. 373–382 (2011)
20. Lerner, B., Elberty, L., Poole, N., Krishnamurthi, S.: Verifying Web Browser Extensions Compliance with Private-Browsing Mode. In: Proceedings of the 18th European Symposium on Research in Computer Security (ESORICS), pp. 57–74 (2013)
21. Mozilla: Jetpack. https://wiki.mozilla.org/Jetpack (2014)
22. Muttis, F., Sacco, A.: HTML5 heap sprays. http://exploiting.files.wordpress.com/2012/10/html5-heap-spray.pdf (2012)
23. Nguyen, N.: Please read: security issue on AMO. http://blog.mozilla.org/addons/2010/02/04/please-read-security-issue-on-amo/ (2010)
24. Ratanaworabhan, P., Livshits, V.B., Zorn, B.G.: Nozzle: a defense against heap-spraying code injection attacks. In: Proceedings of the 18th USENIX Security Symposium, pp. 169–186 (2009)
25. Rieck, K., Krueger, T., Dewald, A.: Cujo: efficient detection and prevention of drive-by-download attacks. In: Proceedings of the 26th Annual Computer Security Applications Conference (ACSAC), pp. 31–39 (2010)

26. Schneier, B.: How the nsa attacks tor/firefox users with QUANTUM and FOXACID. https://www.schneier.com/blog/archives/2013/10/how_the_nsa_att.html (2013)
27. Schwartz, M.: Hackers target Java 6 with security exploits. http://www.informationweek.com/security/vulnerabilities/hackers-target-java-6-with-security-expl/240160443 (2013)
28. Sterne, B., Barth, A.: Content security policy 1.0. W3C Candidate Recommendation (2012)
29. Stone-Gross, B., Abman, R., Kemmerer, R.A., Kruegel, C., Steigerwald, D.G., Vigna, G.: The underground economy of fake antivirus software. In: Proceedings of the 12th Workshop on the Economics of Information Security (WEIS), pp. 55–78 (2013)
30. US-CERT: Oracle Java contains multiple vulnerabilities. Alert (TA13-064A) (2013)
31. Van Acker, S., Nikiforakis, N., Desmet, L., Piessens, F., Joosen, W.: Monkey-in-the-browser: malware and vulnerabilities in augmented browsing script markets. In: Proceedings of the 9th ACM Symposium on Information, Computer and Communications Security (ASIACCS), pp. 525–530. ACM (2014)
32. Younan, Y., Joosen, W., Piessens, F.: Runtime countermeasures for code injection attacks against c and c++ programs. ACM Comput. Surv. **44**(3), 17 (2012)

Chapter 10
Improving Client-Side Web Security

In previous chapters of this book, we explained the importance of Web security in general, and more specifically, client-side Web security. We have presented several threat models, each with different capabilities, and have extensively discussed how these attackers threaten the security of Web applications. We have given an overview of the relevant mitigation techniques and highlighted the current state-of-the-art research results. Finally, we have provided details on the current state of practice and formulated best practices to defend Web applications against numerous attacks.

This chapter summarizes the best practices covered earlier in this book, and boils them down to a "must-have" list of security technologies of the modern age. Additionally, we discuss the role of research in client-side Web security, and identify important areas for future research.

10.1 Overview of Best Practices

As most Web security issues are not new, numerous mitigation techniques have been proposed and many of them are supported by mainstream browsers. Unfortunately, the Web has always suffered from legacy software with a slow update cycle, even for extremely critical vulnerabilities. For example, the server-side Heartbleed vulnerability [32], which is considered to be one of the worst Web problems ever, has been patched almost immediately, and as of this writing, 4 months after its disclosure, SSL Pulse [27] still reports 777 popular sites to be vulnerable. A similar story goes for the *HttpOnly* cookie flag, an effective countermeasure with virtually no impact on a Web application, which only sees a 54 % adoption rate among the Alexa top 10,000 sites, 12 years after its introduction.

In order to improve the current state of practice, we give an overview of the most important best practices, which are essential for improving the security of modern Web applications. All of these technologies are widely supported, as can be verified using the helpful *Can I Use* site [7]. While many of the techniques covered below and explained in detail in this book, are applicable for both new and legacy applications; deploying them for legacy applications may be more challenging.

© Philippe De Ryck, Lieven Desmet, Frank Piessens, Martin Johns 2014   105
P. De Ryck et al., *Primer on Client-Side Web Security*,
SpringerBriefs in Computer Science, DOI 10.1007/978-3-319-12226-7_10

10.1.1 Secure Communication Channel

The lack of a secure communication channel is an enabling factor for numerous other attacks, such as session hijacking, compromising script inclusions, etc. Therefore, the most important best practice is deploying Web applications over a properly configured Transport Layer Security (TLS) channel, a measure not only useful for new Web applications but also for legacy applications. Several resources offer detailed insights into a proper TLS configuration [28, 29, 33], of which the following attention points are most relevant for Web applications:

• Deploy the latest version of TLS, using good cipher suites that offer perfect forward secrecy.
• Avoid using mixed Hypertext Transfer Protocol (HTTP) and Hypertext Transfer Protocol Secure (HTTPS) content, as HTTP content is easily manipulated on the network.
• Use *Strict Transport Security* [12] for HTTPS-only deployments, to prevent a potentially forged HTTP request from ever leaving the browser.
• Mark all cookies that are used over an HTTPS connection as *Secure*, to prevent cookies from being leaked over (forged) HTTP requests.

A very useful tool for verifying the configuration of a TLS deployment is Qualys' SSL Labs Web site [26], which checks your deployment for common vulnerabilities, insecure ciphers, and misconfiguration.

The technologies listed above are currently available in modern browsers, but several promising technologies are still in active development and are worth keeping track of. The most promising technologies aim to address the forging of TLS certificates, using either public key pinning [9], or by using DNSSEC records to certify the key associated with the certificate (DANE) [13].

10.1.2 Application-level Techniques

Many of the attacks discussed in previous chapters can be mitigated on the application level. These mitigation techniques are often supported by popular Web application frameworks or offered by useful libraries. We identify two classes of techniques: design-level techniques that prevent vulnerabilities by design and locally applicable code-level techniques that actively mitigate specific attacks.

On the design level, an important best practice is the use of a multifactor authentication system. Such systems significantly increase the security of the authentication process, largely mitigating phishing scams, brute-forcing attacks, or the theft of credentials. By using an authentication provider, multifactor authentication can be integrated into your own authentication process, or the whole authentication process can be outsourced. Additionally, several well-known providers allow traditional authentication on trusted machines and only enable multifactor authentication in other scenarios. A second, related, design-level best practice is to protect sensitive operations with a reauthentication request. This practice prevents a user from performing unintended operations, for example when misdirected in a clickjacking attack.

On the code level, a developer can take several countermeasures to tighten an application's security, effectively mitigating many attack scenarios. We give a brief overview of the most common code-level countermeasures that should be applied in any Web application:

- Context-sensitive sanitization of outputs is crucial in preventing injection vulnerabilities. This should be the first line of defense against cross-site scripting and scriptless injection attacks, potentially supplemented with a strict Content Security Policy (CSP), as discussed in the next section.
- In order to prevent forged requests, token-based approaches are an effective mitigation technique. Every sensitive operation should be authorized by a token, to ensure its authenticity. Additionally, Web applications should reject cross-origin requests when they are unexpected, which can be checked using the `Origin` header.
- By renewing the session identifier after a change in privilege, session fixation attacks can be effectively mitigated, and the scope of session hijacking attacks can be limited.
- Web application developers should be aware that when they include third-party scripts, they implicitly trust the third party to be non-malicious, and remain free of compromise. The risk of a compromise of a third party automatically spreading to your Web application can be reduced by placing the third-party code within the origin of the Web application.

10.1.3 Security Policies

Server-driven, browser-enforced policies inform the browser about the application's behavior, enabling the browser to block any deviating action, which are potentially malicious. As a best practice, we recommend the use of three widely supported policies, discussed below: the *HttpOnly* restriction on cookies, the use of a strict framing policy, and the use of a strict CSP.

Every cookie issued by a Web application, that is not used by JavaScript within the browser, should be flagged as *HttpOnly*. This applies to most cookies issued today, and should especially be true for cookies holding sensitive tokens, such as session identifiers or authentication tokens.

Framing policies restrict the origins that are allowed to frame the application that defines the policy. By doing so, an application can prevent framing by a malicious Web page, which may be trying to misdirect the user, for example using a clickjacking attack. Modern browsers support two framing policies, the *X-Frame-Options* policy [31] and the *frame-ancestors* directive in CSP [36], of which the latter is the more expressive. Restricting the set of origins that is allowed to frame a page may not always be possible. In that case, the application should restrict framing on all possible pages and ensure that only non-sensitive pages can be framed by any origin. In the near future, the upcoming UI Security specification [20] will offer fine-grained heuristics to determine the legitimacy of the user's interactions.

CSP [36] mainly aims at preventing actions triggered by an attacker who injects content into the application page, of which cross-site scripting is a well-known example. A CSP policy is not meant as a primary defense mechanism against injection attacks but merely aims at restraining an attacker that manages to break through the existing injection defenses. To effectively prevent injection attacks, CSP needs to disable inline scripting, a practice many applications depend on, for example, when defining JavaScript handlers in attributes. Due to this dependency, CSP may be less suited to retrofit to legacy applications but is certainly a viable option for newly developed applications, which can take this into account. Additionally, the upcoming version of CSP [2] will support script nonces, which allow predefined script blocks to be executed, even when placed inline.

10.2 Research-driven Security Technology

Many of the technologies recommended above as a best practice and discussed earlier in this book have resulted from security research. These technologies are an important valorization and dissemination trajectory for research results as they are adopted by mainstream browsers and are essentially deployed on almost all Web-connected machines throughout the world. Finding the right synergy between research results and mainstream is not trivial. Success stories are CSP [35] and Strict Transport Security [16], which went from research proposal to deployment in about a year. On the other side, research results can take several years before being picked up [4] or do not make it at all, as illustrated by the numerous proposals for improving session management [3, 5, 6, 11, 24].

On the other hand, research on currently adopted mechanisms is important to determine the feasibility of certain techniques, especially when deploying them for legacy applications. One example is research on the use of CSP [41], providing insights in the shortcomings of CSP for legacy applications, which in turn drives the next version of the specification [2].

Apart from determining the impact of current security technologies on legacy applications, other research areas are also worth exploring. One ongoing research problem is the integration of potentially untrusted JavaScript into a Web application. Numerous proposals have been made in the past 6 years [1, 15, 19, 21–23, 25, 38, 40], but as of this writing, there is no practical solution ready for deployment. Bringing these valid but often complex proposals towards the modal developer is crucial for ensuring adoption.

Similar to Web technologies and security mechanisms, research is shifting towards the client side. Recent papers aim to detect vulnerable Web sites in the browser [34], focus on vulnerabilities that only exist at the client side, such as DOM-based cross-site scripting [37] or exploits of new HTML5 APIs [39], and investigate the security of browser extensions [17].

Finally, as TLS becomes more important every day, it receives a significant amount of focus from the research community. Research does not only focus on the crypto-graphical properties of TLS [8, 10, 30] but also investigates current deployments [14] and proposes countermeasures to prevent attacks such as man-in-the-middle [18]. As TLS currently offers an all-or-nothing solution, cutting out any intermediaries out of the communication channel, interesting research challenges lie in the controlled integration of these intermediaries. Example scenarios are enabling Web caches when using TLS, allowing certain parties to embed content in designated parts of Web pages, and allowing perimeter security solutions to inspect TLS traffic.

10.3 Conclusion

A result from the evolution towards client-enforced security policies is that we now have multiple defensive technologies against specific Web attacks, enabling a defense-in-depth strategy. For example, by deploying several mitigations against cross-site scripting attacks, the harm of a successful cross-site scripting exploit can be severely limited, or even prevented altogether. As the complexity of Web applications grows, legacy systems will need to be protected, such defense-in-depth strategies will become increasingly important.

A final conclusion to draw from this book is that Web security is a continuous race between attackers and defenders, similar to the security of other complex systems. On one hand, we see regular discoveries and disclosures of new attacks, on the other hand, we have a strong research community working on new defenses, as well as security-aware browser vendors incorporating state-of-the-art technologies. Due to this fast pace, it is more important than ever to stay up-to-date with the latest technology, which is precisely the goal of this book.

References

1. Agten, P., Van Acker, S., Brondsema, Y., Phung, P.H., Desmet, L., Piessens, F.: JSand: complete client-side sandboxing of third-party JavaScript without browser modifications. In: Proceedings of the 28th Annual Computer Security Applications Conference (ACSAC), pp. 1–10 (2012)
2. Barth, A., Veditz, D., West, M.: Content security policy level 2. W3C Working Draft (2014)
3. Bortz, A., Barth, A., Czeskis, A.: Origin cookies: session integrity for Web applications. Web 2.0 Security and Privacy (W2SP) (2011)
4. Chen, E.Y., Bau, J., Reis, C., Barth, A., Jackson, C.: App isolation: get the security of multiple browsers with just one. In: Proceedings of the 18th ACM Conference on Computer and Communications Security (CCS), pp. 227–238 (2011)
5. Dacosta, I., Chakradeo, S., Ahamad, M., Traynor, P.: One-time cookies: preventing session hijacking attacks with stateless authentication tokens. ACM Trans. Internet Technol. (TOIT) **12**(1), 31 (2012)

6. De Ryck, P., Desmet, L., Piessens, F., Joosen, W.: Eradicating bearer tokens for session management. W3C/IAB Workshop on Strengthening the Internet Against Pervasive Monitoring (STRINT) (2014)
7. Deveria, A.: Can i use ... support tables for HTML5, CSS3, etc. http://caniuse.com (2014)
8. Duong, T., Rizzo, J.: BEAST - Here Come The XOR Ninjas. http://nerdoholic.org/uploads/dergln/beast_part2/ssl_jun21.pdf (2011)
9. Evans, C., Palmer, C., Sleevi, R.: Public Key Pinning Extension for HTTP. IETF Internet Draft (2014)
10. Gluck, Y., Harris, N., Prado, A.: BREACH: reviving the CRIME Attack. http://breachattack.com/resources/BREACH%20-%20SSL,%20gone%20in%2030%20seconds.pdf (2013)
11. Hallam-Baker, P.: Http integrity header. IETF Internet Draft (2012)
12. Hodges, J., Jackson, C., Barth, A.: HTTP strict transport security (HSTS). RFC Proposed Standard (RFC 6797) (2012)
13. Hoffman, P., Schlyter, J.: The DNS-based authentication of named entities (DANE) transport layer security (TLS) protocol: TLSA. RFC Proposed Standard (RFC 6698) (2012)
14. Huang, L.S., Rice, A., Ellingsen, E., Jackson, C.: Analyzing forged ssl certificates in the wild. In: Proceedings of the 35th IEEE Symposium on Security and Privacy (SP) (2014)
15. Ingram, L., Walfish, M.: Treehouse: Javascript sandboxes to help web developers help themselves. In: Proceedings of the USENIX annual technical conference (ATC) (2012)
16. Jackson, C., Barth, A.: ForceHTTPS: protecting high-security web sites from network attacks. In: Proceedings of the 17th international conference on World Wide Web (WWW), pp. 525–534 (2008)
17. Kapravelos, A., Grier, C., Chachra, N., Kruegel, C., Vigna, G., Paxson, V.: Hulk: eliciting malicious behavior in browser extensions. In: Proceedings of the 23rd USENIX Security Symposium, pp. 641–654 (2014)
18. Karapanos, N., Capkun, S.: On the effective prevention of tls man-in-the-middle attacks in Web applications. In: Proceedings of the 23rd USENIX Security Symposium, pp. 671–686 (2014)
19. Magazinius, J., Phung, P.H., Sands, D.: Safe wrappers and sane policies for self protecting javascript. In: Proceedings of the 15th Nordic Conference on Secure IT Systems (NordSec), pp. 239–255 (2010)
20. Maone, G., Huang, D.L.S., Gondrom, T., Hill, B.: User interface safety directives for content security policy. W3C Last Call Working Draft (2014)
21. Meyerovich, L., Livshits, B.: ConScript: specifying and enforcing fine-grained security policies for Javascript in the browser. In: Proceedings of the 31st IEEE Symposium on Security and Privacy (SP), pp. 481–496 (2010)
22. Mickens, J.: Pivot: fast, synchronous mashup isolation using generator chains. In: Proceedings of the 35th IEEE Symposium on Security and Privacy (SP), pp. 261–275 (2014)
23. Miller, M.S., Samuel, M., Laurie, B., Awad, I., Stay, M.: Caja: safe active content in sanitized javascript. http://google-caja.googlecode.com/files/caja-spec-2008-01-15. pdf (2008)
24. Murdoch, S.J.: Hardened stateless session cookies. Security Protocols XVI, pp. 93–101 (2011)
25. Phung, P.H., Sands, D., Chudnov, A.: Lightweight self-protecting javascript. In: Proceedings of the 4th ACM Symposium on Information, Computer and Communications Security (ASIACCS), pp. 47–60 (2009)
26. Qualys: Qualys SSL labs. https://www.ssllabs.com/ (2014)
27. Qualys: Trustworthy internet movement—ssl pulse. https://www.trustworthyinternet.org/ssl-pulse/ (2014)
28. Ristić, I.: OpenSSL cookbook. Feisty Duck (2013)
29. Ristić, I.: Bulletproof SSL and TLS. Feisty Duck (2014)
30. Rizzo, J., Duong, T.: The CRIME attack. https://docs.google.com/presentation/d/11eBmGiHb-YcHR9gL5nDyZChu_-lCa2GizeuOfaLU2HOU/edit?pli=1#slide=id.g1d134dff_1_222 (2012)
31. Ross, D., Gondrom, T.: HTTP Header Field X-Frame-Options. RFC Informational (RFC 7034) (2013)

32. Schneier, B.: Hearbleed. https://www.schneier.com/blog/archives/2014/04/heartbleed.html (2014)
33. Sheffer, Y., Holz, R., Saint-Andre, P.: Recommendations for Secure Use of TLS and DTLS. IETF Internet Draft (2014)
34. Soska, K., Christin, N.: Automatically detecting vulnerable websites before they turn malicious. In: Proceedings of the 23rd USENIX Security Symposium, pp. 625–640 (2014)
35. Stamm, S., Sterne, B., Markham, G.: Reining in the web with content security policy. In: Proceedings of the 19th international conference on World wide web (WWW), pp. 921–930 (2010)
36. Sterne, B., Barth, A.: Content security policy 1.0. W3C Candidate Recommendation (2012)
37. Stock, B., Lekies, S., Mueller, T., Spiegel, P., Johns, M.: Precise client-side protection against dom-based cross-site scripting. In: Proceedings of the 23rd USENIX Security Symposium, pp. 655–670 (2014)
38. Ter Louw, M., Ganesh, K.T., Venkatakrishnan, V.: AdJail: practical Enforcement of Confidentiality and Integrity Policies on Web Advertisements. In: Proceedings of the 19th USENIX Security Symposium, pp. 371–388 (2010)
39. Tian, Y., Liu, Y.C., Bhosale, A., Huang, L.S., Tague, P., Jackson, C.: All your screens are belong to us: attacks exploiting the html5 screen sharing api. In: Proceedings of the 35th IEEE Symposium on Security and Privacy (SP), pp. 34–48 (2014)
40. Van Acker, S., De Ryck, P., Desmet, L., Piessens, F., Joosen, W.: WebJail: least-privilege integration of third-party components in web mashups. In: Proceedings of the 27th Annual Computer Security Applications Conference (ACSAC), pp. 307–316 (2011)
41. Weinberger, J., Barth, A., Song, D.: Towards client-side html security policies. In: Proceedings of the 6th USENIX Workshop on Hot Topics on Security (HotSec) (2011)